DIGITAL DETOX

THE ULTIMATE GUIDE TO BEATING
TECHNOLOGY ADDICTION, CULTIVATING
MINDFULNESS, AND ENJOYING MORE
CREATIVITY, INSPIRATION, AND BALANCE IN
YOUR LIFE!

DAMON ZAHARIADES

ARTOFPRODUCTIVITY.COM

CONTENTS

PART II
10 STEPS TO DOING A COMPLETE DIGITAL DETOX

PART III
THE EFFECTS OF A DIGITAL DETOX ON YOUR BRAIN

OTHER BOOKS BY DAMON ZAHARIADES

The Mental Toughness Handbook

The definitive, step-by-step guide to developing mental toughness! Exercises included!

To-Do List Formula

Finally! Discover how to create to-do lists that work!

The Art Of Saying NO

Are you fed up with people taking you for granted? Learn how to set boundaries, stand your ground, and inspire others' respect in the process!

The Procrastination Cure

Discover how to take quick action, make fast decisions, and finally overcome your inner procrastinator!

Fast Focus

Here's a proven system that'll help you to ignore distractions, develop laser-sharp focus, and skyrocket your productivity!

The 30-Day Productivity Plan

Need a daily action plan to boost your productivity? This 30-

day guide is the solution to your time management woes!

The 30-Day Productivity Plan - VOLUME II

30 MORE bad habits that are sabotaging your time management - and how to overcome them one day at a time!

The Time Chunking Method

It's one of the most popular time management strategies used today. Triple your productivity with this easy 10-step system.

80/20 Your Life!

Achieve more, create more, and enjoy more success. How to get more done with less effort and change your life in the process!

Small Habits Revolution

Change your habits to transform your life. Use this simple, effective strategy for adopting any new habit you desire!

Morning Makeover

Imagine waking up excited, energized, and full of self-confidence. Here's how to create morning routines that lead to explosive success!

The Joy Of Imperfection

Finally beat perfectionism, silence your inner critic, and overcome your fear of failure!

The P.R.I.M.E.R. Goal Setting Method

An elegant 6-step system for achieving extraordinary results in every area of your life!

For a complete list, please visit

http://artofproductivity.com/my-books/

YOUR FREE GIFT

~

I have something for you. It won't cost you a dime. It's a 40-page guide titled *Catapult Your Productivity! The Top 10 Habits You Must Develop To Get More Things Done.* I'd like you to have a free copy with my compliments.

You can grab your copy by clicking on the following link and joining my mailing list:

http://artofproductivity.com/free-gift/

Before we jump into *Digital Detox*, I'd like to thank you. You're taking a chance on me with your time and money. I appreciate it. Giving you a free PDF copy of my guide *Catapult Your Productivity* is my way of showing my appreciation.

On that note, let's dig in.

FOREWORD

~

This action guide is made up of two main parts. The first part deals with the psychology behind technology addiction. The second part addresses how to do a digital detox.

I'm a firm believer that the best way to resolve a behavioral problem is to completely understand it. That includes identifying its root causes, appreciating its short and long-term effects, and comprehending the many ways in which overcoming the problem will affect the individual's life.

In my opinion, the wrong approach is to gloss over these details and immediately apply draconian measures intended to effect a particular outcome. I believe it's unproductive and futile. For example, demanding that a drug addict stop using drugs will never be effective as a long-term solution. Nor will hiding his drugs, flushing them down the toilet, or burning them.

The compulsion to use will persist. And that means the addiction will rear its head at the first opportunity.

For that reason, I believe it's imperative to understand every facet of technology addiction if we hope to beat it.

This action guide has been written with that philosophy in mind. We'll first define technology addiction to ensure we're starting from the same place. We'll then take a look at the most common objects of obsession among tech junkies. We'll also discuss the telltale signs of addiction as well as the side effects most likely to diminish your quality of life.

We'll talk about the reasons your phone addiction or internet addiction may have a stranglehold on you. And we'll cover how taking a technology sabbatical - i.e. doing a digital detox - will improve your life, both now and in the future.

I've given considerable space in this action guide to the above items. I think they're essential to the process of beating your tech addiction. Once you fully understand the problem, you'll be armed with the information you need to overcome it.

Let's get started...

WHAT IS TECHNOLOGY ADDICTION?

∽

Tell me if this sounds familiar.

You wake up in the morning and immediately reach for your phone. You check for new emails and texts despite having no urgency to do so.

Satisfied that you haven't missed any important messages, you log onto Facebook. You check to see whether your friends posted any updates while you were sleeping.

Then it's off to Twitter. You follow hundreds of friends, acquaintances, and celebrities. Surely *someone* has posted something worth reading. Unfortunately, they haven't.

With nothing of interest happening on social media, you visit your favorite news media sites to find out what's going on the world. After all, you don't want to miss the day's major headlines.

And so the morning progresses. From the moment you

wake up until you've thrown the first cup of coffee down your throat, your attention is dominated by technology. Texts, emails, social media, games, news headlines, blogs, and YouTube videos hold you captive in a vice-like grip.

Worse, the rest of the day follows the same course. Your phone buzzes, signaling the arrival of a new text and you find yourself unable to resist checking it. You receive a notification in your browser that a new email has arrived and you immediately drop everything to read it. You visit Facebook, promising yourself that you'll only spend a few minutes, only to surf aimlessly for an hour.

If you relate to the above circumstances, I have bad news. You're likely addicted to technology. The good news is that you can beat the addiction and reclaim your life. This action guide will show you how.

TECHNOLOGY ADDICTION DEFINED

~

A ll addictions, from an obsession with gambling and drugs to video games and technology, are based on the same fundamental dynamic: the brain's expectation that engaging in a particular activity will produce a reward. The reward may not be obvious to the addict. In fact, it's sometimes counterintuitive since it poses potential harm. But the brain still interprets it as a positive experience.

For example, consider gambling. Most people assume individuals with a gambling addiction are compelled to continue to gamble because they occasionally win their bets. The man at the blackjack table stays routed to his seat because he occasionally beats the dealer's hand. The woman at the craps table sticks around because she wins

her pass line bet and a few place bets before the shooter
rolls a seven.

In reality, although the "high" of winning is a moti-
vating factor for the problem gambler, it's not the main
reward - at least, not to the brain's complex reward system.
The rewarding stimuli - the thing that spurs gamblers to
continue gambling - is the risk involved in the activity. That
is, winning a $1 bet in blackjack brings little satisfaction.
Risking $100 on a single hand is more gratifying.

Unfortunately, over time, addicts need greater amounts
of the rewarding stimuli to produce the same level of
reward. The pathological gambler who starts feeding his
addiction with $5 bets will eventually graduate to $100 bets
- and more if he has the funds to support his growing
habit. The greater the risk, the more dopamine that is
released in the brain and the greater the resulting sense of
satisfaction.

Let's consider that in the context of an addiction to
technology. It works in the same way as a gambling problem. You're compelled to act by the brain's expectation of
a reward for doing so. Researchers are uncertain regarding
what the brain considers to be rewarding stimuli in the
activity. But studies show that the act of checking emails,
texts, and social media releases dopamine in the same
fashion as wagering large sums in a casino.

The dopamine keeps people hooked.

There's another concern with technology addiction.
Researchers have found that tech addicts are more inclined

to feed their addiction because the gadgets that make it possible are everywhere. They're always within reach.

Nearly everyone has a smartphone (Pew Internet found that 68% of adults in the U.S. owned at least one). Tablet ownership is nearly as prevalent; 45% of U.S. adults own one. And of course, laptops and desktop computers are as common as dirt. The tools needed to feed your technology addiction are ubiquitous. They're inescapable.

The result? The compulsion to check your email and texts is difficult to resist. The impulse to check social media is irrepressible. The urge to check your voicemail, seek out the latest news headlines, and visit your favorite blogs and forums is overwhelming.

These are the reasons it's so easy to develop an addiction to technology. First, you have the necessary tools at your disposal 24 hours a day. Second, every time you use them, you stimulate your brain's reward system. Over and over and over again.

Under these circumstances, dependency and addiction are practically foregone conclusions.

COMMON "DRUGS" OF CHOICE FOR THE TECHNOLOGY ADDICT

~

Technology spans a variety of devices and platforms. From smartphones, tablets, and video game consoles to social media, news media, and email. It's worth addressing how each of these can slowly become an obsession. That way, you'll be better able to recognize whether you have a problem. (We'll talk about the telltale signs of addiction in the next chapter.)

Smartphones

When it comes to tech-related devices, this is the big one. Smartphones have become so pervasive in our society that most of us now carry one. The problem is, their use stimulates the reward center of the brain in a way that encourages dependency. This dependency sets the stage for the onset of a full-blown addiction.

Our phones are tools of compulsion. We receive a text and instantly read it and respond. We receive a phone call and immediately answer it. Between the texts and calls, we neurotically check Facebook for new posts and Twitter for new tweets. When we're done with social media, we check our email, search for news headlines, and watch YouTube videos.

In short, we're addicted.

You've probably seen people sitting together at restaurants looking at their phones instead of interacting with each other. Maybe you're one of them.

Don't beat yourself up over it. Smartphone addiction is surprisingly common. Worse, it's easy to develop. I'm going to show you how to break the habit in this action guide.

Tablets

Tablets like the iPad, Google Pixel and Samsung Galaxy Tab are becoming nearly as prevalent as smartphones. More people than ever are carrying these devices wherever they go (in addition to their phones). They're using them to go online, check their email, watch videos, reply to texts, and search for news and other "important" content.

That sounds good at first. After all, who wouldn't like the ability to obtain information, connect with friends and loved ones, and enjoy a variety of entertainment at a moment's notice?

Of course, the problem is that it's easy to become addicted. It's like eating chocolate. An occasional piece can be a wonderful treat. But it's easy to overindulge, develop a dependency, and form an addiction.

Computers

We usually think of computers as tools that help us to get things done. For example, some of us create spreadsheets for our jobs. Others among us develop software or apps. Still others use them primarily as word processors. In fact, I'm on my laptop at this moment typing this action guide.

In addition to being a productivity tool, computers are tools of comfort. When we're stressed, we use them to relax (e.g. play solitaire). When we're bored, we use them to entertain ourselves (e.g. watch YouTube videos). When we're procrastinating, we use them to distract our attention (e.g. check Facebook).

Over time, our computers can become crutches. We begin to rely on them to give us a sense of fulfillment whenever we're feeling anxious, bored, or depressed. That's when we run the risk of allowing them to become a compulsion.

The Internet

We spend a lot of time on the internet. Researchers report that our consumption of online media doubled between 2010 and 2015.

And that trend is unlikely to change anytime soon.

As you probably know from experience, much of our time spent online is wasted. When we're on the internet, we're not always researching things for our jobs or moving projects forward. Instead, we're checking our email every 20 minutes, watching YouTube videos, and surfing social media. We're playing games, reading up on current events, and visiting our favorite forums.

These activities seem harmless on the surface. But the quick and easy accessibility of the internet, along with the dopamine rush people experience when they go online, makes them dangerous. They become habits, which can lead to compulsion, dependency, and addiction.

Internet addiction is a growing concern among psychologists and psychiatrists. The number of diagnosed cases grows each year. In fact, many experts believe the condition should be added to the *Diagnostic and Statistical Manual of Mental Disorders*.

Video Games

Gaming has been one of the chief obsessions among young people over the last 20 years. Long ago, you needed a dedicated home console, such as the Atari 2600, Super Nintendo, X-Box, or Playstation. Not so today. These days, people play games on their phones, through their internet browsers, and on handheld game consoles like the Nintendo 3DS and Playstation Vita.

There are more opportunities to play, and thus more opportunities to develop an obsession.

I can relate to this. I used to spend hours each and every night playing video games. Doing so made me feel good. It was a way to escape the day's stresses. But in retrospect, it was a terrible way to spend my time since it yielded no tangible, lasting benefits. I have nothing to show for the time I spent playing the games.

The compulsion to play video games can become all-consuming. Some addicts are so obsessed that they wear adult diapers so they can play without interruptions. I'm happy to say my gaming addiction never reached that point. But it reveals how serious this obsession can become in some people.

Social Media

Social media is arguably one of the most insidious "drugs" of choice among technology addicts. In the same way illicit narcotics like heroine are designed to be addictive, so too are social media websites like Facebook. The engineers who designed the sites have done everything in their power to make sure you're compelled to visit them over and over.

Have you ever reached for your phone right after waking up in the morning to check for Facebook updates? Ever pulled your phone out at a restaurant to read the latest tweets on Twitter? Ever log onto Pinterest, Instagram, or Google Plus multiple times during the workday, looking over your shoulder to make sure your boss doesn't notice?

If so, you might be addicted to social media.

The good news is that you're not alone. Millions of people experience - and act upon - the same compulsions every single day.

The bad news is that the obsession with social media is almost certainly producing unhealthy side effects. Scientists have discovered that constant exposure to websites like Facebook and Twitter can alter the brain, affecting the ability to process emotions. It can also lead to restlessness, negative self-image, a decline in happiness, and in extreme cases, depression.

And that doesn't even address the negative effect of social media addiction on your productivity and relationships.

News Media

News can be extremely addictive. All of us want to feel informed about current events. We want to be kept abreast about what is happening in the world.

Unfortunately, that inclination can easily turn into an obsession because of today's rapid-fire news cycle. We no longer have to wait for the morning newspaper or our favorite nightly news program on TV to find out about the latest headlines. Instead, we can go online and discover the latest events and happenings on a minute-by-minute basis.

That's a form of nirvana for the news addict.

I speak from experience. There was a time in my life when I was addicted to headlines. While sitting in my office, I'd visit CNN, the New York Times, and the Drudge Report every five minutes. I had to know what was happening in the world at all times. You can imagine the outcome. The constant distraction prevented me from getting anything done.

The continuous barrage of updated headlines is made worse by news media sites' attempts to attract eyeballs. The more traffic they receive, the more ads they can display. That translates into increased revenue. So they lead with tragedy and make every headline seem significant.

Psychologists say our brains feed on negative triggers. We're hardwired that way. The negativity causes us to become fearful and stressed, feeding our desire for "news" and eventually creating an obsession. Websites like CNN.com and NYTimes.com are aware of this cognitive response. They leverage it to their benefit.

Email

Do you check your email more than once per hour? Or worse, do you leave a browser tab open so you can see new emails as they come in? Do you grab your phone whenever you receive a notification that someone has sent you an email?

If so, you're not alone.

Millions of people are compelled to check their email dozens of times a day. It's a seductive impulse. Checking their email stimulates the reward centers of their brains. The stimulation triggers the release of dopamine, which brings them pleasure.

That's the same process that feeds addictions to illicit drugs. It's no wonder so many people are addicted to email.

According to a doctor at the Center for Internet Studies, there are approximately 11 million "email junkies" out there. These individuals might regularly attend their kids' baseball games, piano recitals, and other functions. They might go on weekly dates with their spouses. They might even go on a few vacations per year. But no matter where they are, their minds are always on their email.

Our obsession with email is just as common as our obsession with social media. Part of the problem is that we possess the tools (our phones) to check for new messages whenever we desire. In fact, we can program our phones to notify us the moment new messages arrive.

These notifications, combined with our inability to ignore them and unwillingness to turn them off, ensure our continued addiction.

Blogs and Forums

Blogs and forums are dangerous for habitual procrastinators. They can easily become a constant source of distraction, similar to social media and email. To that end, they're among the most common productivity killers.

Consider blogs.

We read them because we find them to be interesting or entertaining, or both. Some blogs are so popular that each post attracts dozens, and sometimes hundreds, of comments. The same commenters show up over and over, forming mini-communities where everyone knows each other.

Forums fill many of the same needs.

Some of us frequent them to keep abreast of the latest developments in our fields of expertise or areas of interest. Sometimes, we visit for the entertainment value. Regardless of our purpose, most of us enjoy interacting with people who share our interests. Forums are online communities filled with like-minded individuals.

The danger is that these communities, found both on blogs and forums, can serve as distractions that hamper our productivity. If we're working on a task or project that's difficult or unappealing, we're more likely to succumb to the distraction. That prevents us from getting things done.

Worse, the more often we do so, the more doing so becomes a habit. And that's the first step toward addiction.

Some people spend several hours each day on forums. I know from firsthand experience because I used to be one of them.

YouTube

Videos are as much a part of the online experience as social media and email. According to the Pew Research Center, seven out of ten adults have used YouTube or Vimeo at some point. More than 30% have posted videos online.

A lot of folks visit YouTube to learn new skills. For example, I've lost count of the number of cooking videos I've watched over the years. Others tune in for entertainment; they watch music videos, cat videos, and clips of the week's funniest "fails."

The problem is, it can become habit-forming for three reasons.

First, most videos are relatively short (under 10 minutes), giving the impression that watching them won't require much time. Of course, few people watch just one video; they watch several in a row.

Second, YouTube encourages users to share their favorite videos with their friends. This social aspect is what makes websites like Facebook and Twitter so addictive.

Third, a flood of new content is uploaded every day. According to Pew Research, 100 hours of video are uploaded to YouTube *each minute*. There's always something new to watch.

Because we can access YouTube through our phones, informative and educational content is always at our fingertips. We can search for, and play, an endless stream of videos on a whim.

That's dangerous. It's easy to waste countless hours that could otherwise be spent doing something more rewarding.

Again, I speak from experience.

Let's now shift gears and talk about the most common signs of technology addiction. If you can relate to any of those described in the next chapter, it's time to do a digital detox.

10 SIGNS YOU MAY BE ADDICTED TO TECHNOLOGY

∿

A ddiction leaves clues. It manifests in ways that are easy to recognize if you know what to look for.

Many clues vary according to the tool feeding your addiction. For example, if you're addicted to your smartphone, you'll notice symptoms that are different than those experienced by a video game addict.

Having said that, some signs of technology addiction are similar across the board. One example is restlessness, a common trait that presents in people forced to go without their "drugs" of choice for extended periods.

Below, you'll learn the ten most common signs of tech obsession. If you recognize one or two in your life, that doesn't necessarily mean you have an addiction. But if you experience several on a regular basis, it's time to go on a digital detox.

#1 - You instinctively reach for your phone whenever it rings or buzzes.

It's one thing to expect an important call or email, and check your phone out of a sense of true urgency. It's another thing entirely to compulsively grab your phone whenever it vibrates or makes a sound.

This reaction suggests you've developed a habit. You're responding automatically to a trigger, similar to Pavlov's salivating dogs.

#2 - You become anxious if you don't check your phone after receiving an alert.

Imagine you're having a conversation with a friend. Your phone, hidden in your pocket, vibrates with a notification of an incoming message. You don't know whether you've received a text, email, or phone call. It might even be a trivial alert from one of your apps.

Do you feel a twinge of angst if you're unable to check your phone? Does the feeling intensify until you're finally able to do so? Or worse, do you find yourself constantly checking your phone during the conversation on the slight chance you missed an email, text, or call?

This proclivity points to a compulsion. The longer

you're prevented from checking your phone, the more anxiety you feel.

#3 - You experience withdrawal symptoms if you can't get online.

Internet addiction is real. Millions of people suffer from it. And when they're prevented from going online, they experience withdrawal symptoms that are like those suffered by drug addicts.

A 2013 study in the UK found that "high internet users" became moodier and more depressed than "low internet users" when they were forced to stop surfing the net. The researchers reported that the study's volunteers suffered a "comedown" that was similar to the comedown seen after a drug high.

Do you feel irritable when you're unable to get online? Do you become shaky when you're forced to endure long periods without internet access? Do you get defensive when people ask about your surfing habits? Has your social life deteriorated as you spend more time online?

If so, there's a good chance you're addicted.

#4 - You're habitually late or routinely fail to meet your commitments because of technology.

Technology *can* make you more punctual. Thousands of apps are available for your phone that will help you to manage your calendar and schedule. Thousands of extensions are available for your internet browser for the same purpose.

But these tools offer little help for the technology addict. Using them productively requires diligence and discipline, traits that are difficult for the addict to maintain for long periods. Instead, the addict's obsession with his phone, the internet, video games, news headlines, YouTube, and social media negatively affect his punctuality. They make him late, causing others to consider him unreliable.

Are you frequently late for school, work, or appointments because you're preoccupied with your gadgets? Does your fixation with your phone or the internet cause you to break commitments - or worse, simply not show up when others expect you to?

These are signs of addiction.

#5 - You feel euphoric when you check social media.

You know the feeling. You get a high, or buzz, whenever you log into Facebook to check out what your friends are up to. You feel a sudden happiness, even excitement, when you surf Twitter, Pinterest, or Instagram.

The effect occurs because dopamine floods the brain's pleasure circuit. It thereby induces euphoria, a response seen with nearly all addictive drugs.

And as you might suspect, every high is followed by a low. A crash. A comedown.

Do you feel inexplicably happy when you log onto your

favorite social media sites? Do you experience a sense of well-being the moment you visit your Facebook or Twitter pages?

If so, your brain's reward system is reacting in the same way as that of an addict.

#6 - You take your phone or tablet with you into the bathroom.

It's one thing to enjoy having reading material available when nature calls. But that's if you're planning to spend at least a few minutes answering the call. If you're taking your phone or tablet with you for a 30-second pit stop, you're probably obsessed.

Do you grab your phone or tablet every time you visit the restroom? Do you use them when visiting public restrooms, where most people are inclined to hurry and finish their business? Do you feel disappointed when you forget to bring your phone with you?

If you answered yes to any of the above, you're an excellent candidate for a digital detox.

#7 - You sacrifice sleep to spend more time online.

Sleep is one of the first things we forfeit when we're busy. This is ironic since getting adequate sleep is the key to functioning well. If you're exhausted, your focus and performance will suffer, whether in the office, on the basketball court, or while having a simple conversation with your spouse.

Technology addiction will impair your sleep. Ask yourself, have you ever stayed up late to watch YouTube videos in bed? Have you ever texted friends or spent time on Facebook long after you knew you should have turned out the lights? Have you ever stayed up past your normal bedtime playing video games?

Of course, answering "yes" to any of the above doesn't automatically mean you're a tech addict. We've all done it. And not all of us are true addicts. But if you regularly sacrifice sleep to surf the internet, play video games, or hang out on social media, you may indeed be a "techaholic."

#8 - You no longer participate in activities you once enjoyed.

When your friends call you to get together, are you inclined to say "no" because you'd rather stay home and play with your phone apps? Do you now spend time on Facebook that you once spent playing basketball, learning the guitar, or hanging out with your kids? Do you text friends while ignoring your spouse, who's sitting next to you?

These are telltale signs of tech addiction. You're obsessed to the point that other activities hold little interest for you. With time, your friends will drift away, your family will feel neglected, and your former hobbies will be all but forgotten.

A digital detox will remind you of the important things in your life you've been sacrificing to feed your addiction.

#9 - Your social skills have deteriorated to the point that you're uncomfortable around others.

Experts claim the internet has levied a considerable social cost. It has eroded the amount of time we spend with our friends and loved ones. In the process, it has changed the way we interact with each other. In many ways, it has impaired our ability to communicate ideas.

These effects are more pronounced for the internet addict. He uses the internet as a form of escapism, minimizing the challenges that accompany social interaction. The more time he spends online, the further his interpersonal skills decline. He becomes less able to maintain eye contact and more likely to mumble when talking to others. He may even avoid face-to-face interaction altogether, preferring the relative anonymity of online interaction.

Ask yourself whether you feel comfortable in social settings. Can you have and enjoy an engaging conversation with another person? Are you able to form connections with people who are standing in front of you?

If not, it's time to set your phone down, turn off your computer, and put away your video game console. In short, it's time for a digital detox.

#10 - Multiple attempts to curtail your use of technology have failed.

Most of us have vices. For some, it's ice cream. For others, it's cigars. Still others have a weakness for gambling, shopping, alcohol, or illicit narcotics.

These vices have one thing in common: they're extremely difficult to quit.

Think about the last time you were on a diet. The unhealthy foods you once enjoyed were probably a constant temptation. You may have even fallen off the wagon multiple times, treating yourself to your favorite dessert.

So it goes with all forms of addiction, including an obsession with your phone, the internet, and social media. If you've ever tried to curtail your usage, but suffered multiple relapses, you're likely an addict. You face a psychological imperative to act.

The good news is that you're in the right place. You've decided to address the problem. We're going to talk about the side effects of technology addiction in the next chapter. It's important to appreciate how this type of obsession negatively affects your life so you'll know what's at stake.

TOP 12 SIDE EFFECTS OF TECHNOLOGY ADDICTION

~

Addiction comes with consequences. These consequences range from physical to psychological. Depending on their nature and severity, they make life less enjoyable.

It's worth underscoring that technology has the ability to improve our lives. It can boost our productivity and increase our effectiveness at everything we do. But it's just a tool. And like any tool, it can wreak havoc if used poorly or thoughtlessly.

Following are the 12 most common side effects seen in tech addicts. Individually, these side effects may seem harmless. But together, they show that continued overindulgence with technology can produce catastrophic results.

#1 - Inability To Concentrate

Research has shown that our average attention span is shorter than that of a goldfish. In 2000, we were able to focus for 12 seconds. Today, our attention drifts after eight seconds.

It's safe to say our phones, tablets, computers, video game consoles, and favorite social media sites are partly to blame. They deliver stimuli in short, powerful bursts. They also encourage us to multitask.

Both erode our ability to focus.

That's for non-addicts. Individuals who are addicted to their phones and other gadgets typically struggle with an even shorter attention span. They're prone to distractions and their ability to concentrate is thus even further impaired.

Have you ever had trouble sitting still while watching a movie? Ever had difficulty reading a book for longer than 20 minutes? There's a good chance your obsession with - and continuous exposure to - technology is the reason.

#2 - Inconsistent Sleep Quality

Staring at your phone, tablet, or computer can make it more difficult to fall asleep at night. And as we noted earlier, lack of restful sleep will have a negative impact on your performance and productivity.

According to researchers at the Harvard Medical School, our devices impair our sleep due to the light wavelength used in the screens. The artificial light, known as "blue light," reduces the secretion of melatonin, a hormone that helps us to get to sleep.

If you've been having trouble sleeping at night, the problem may be your constant usage of your phone and other gadgets. You're not alone. Millions of men and women across all age groups suffer from the same side effect.

#3 - Increased Stress

Technology has put us at the beck and call of others. Twenty years ago, the workday began at 8:00 a.m. and ended at 5:00 p.m. Today, we receive work-related text messages and emails outside that time window. We wake up to text messages and go to bed responding to work emails. It's not uncommon for employees to stay in touch with their bosses and coworkers while they're on vacation.

That pattern leads to stress. The workday never truly ends. We never get to unplug from our work-related responsibilities.

Additionally, we feel compelled to respond immediately to texts and emails from our friends and family members. The quickness of our responses is taken as an indication of how much we care.

Our devices are tools that *should* make our lives better. But for the technology addict, they can become a constant source of anxiety.

#4 - Decline In Social Life

Addiction to the internet, video games, news media, and social media has a negative effect on the addict's social life. He stops responding to phone calls from his family. He declines invitations to go out with his friends. Instead, he stays at home, eyes glued to his computer, phone, and video games.

Consequently, the tech junkie's relationships begin to erode. Friendships start to fray. Loved ones become hurt or irritated by his lack of response. Repeated attempts to connect with him become less frequent and eventually stop altogether.

A related side effect is that the addict becomes less able to relate to others. His obsession with his gadgets and the internet causes his interpersonal skills to deteriorate. He forgets how to carry on conversations and establish connections with people he meets in "real life" (face to face).

If you're feeling isolated and lonely, the cause may be your obsession with your phone and other devices.

#5 - Back And Neck Pain

Visualize someone who is standing with good posture. Her back is straight. Her shoulders are properly aligned. Her head is held straight, tilted neither forward nor to the side.

Now visualize someone standing while staring at her phone. Her back may be slightly slouched. But the real problem is the position of her head. It's probably tilted forward at a 45-degree angle. If she's been staring at the screen for several minutes, her head may be tilted at a 60-degree angle.

The adult head weighs weighs approximately 11 pounds. That's a significant amount of weight. When it's tilted forward, it places stress on the spine. The greater the degree of tilting, the more stress the spine must tolerate. According to spinal experts, a 15-degree forward tilt of the head exerts 27 pounds of pressure on the spine. A 45-degree tilt exerts 49 pounds of pressure. At a 60-degree tilt, your head puts 60 pounds of pressure on your spine.

Many people keep their heads in that position for extended periods while they stare at their phones. The result? Chronic stiffness and pain in the back, neck, and shoulders.

#6 - Weight Gain

If you've recently gained weight, your phone may be the culprit. Or more accurately, your obsession with your phone. That's according to researchers who, in 2013, found that college students who spent an average of 14 hours a day on their phones were markedly less fit than students who spent an average of 90 minutes on their phones.

Study participants who were described as "high-frequency" users tended to lead more sedentary lives. They seldom engaged in exercise. They rarely participated in sports. Instead, they were more likely to be couch potatoes, spending their time surfing the internet, watching television, and playing video games.

Long periods of inactivity typically lead to weight gain and poor health. If you're overweight or unhealthy, your phone, tablet, or computer may be to blame.

#7 - Depression

Technology addiction can trigger feelings of depression in a couple of ways. First, as we noted above, the tech junkie begins to feel isolated as he spends more and more time online. He spends less time nurturing meaningful relationships, which leaves him with few interpersonal connections. This circumstance sets the stage for a growing sense of unhappiness. If he remains in this predicament for weeks on end, he may eventually experience feelings of gloom, helplessness, and hopelessness.

The second way an obsession with technology can cause depression involves the addict's use of his phone. When his friends and family members fail to immediately respond to his texts and emails, he starts to feel unimportant to them.

Scientific research has uncovered evidence of this effect. In 2015, a study in the Journal of Medical Internet Research found that phone usage was directly linked to depressive symptoms. Moreover, the severity of the symptoms increased with the extent of the participants' use.

Have you been feeling depressed lately? If so, consider how many hours per day you spend online compared to how many hours you spend connecting face-to-face with friends and loved ones.

#8 - Vision Problems

It's not a coincidence that many people who spend considerable time in front of their computers wear eyeglasses. Looking at a computer all day causes eye strain. Some individuals even experience pain, often a symptom of a broader malady. The effect is known as computer vision syndrome.

Researchers are split on what causes the strain. Some believe it's a form of repetitive stress injury. The eyes move back and forth, hour after hour, an activity that causes them to tire. Others claim it stems from continuous exposure to the ultraviolet light emitted by computer screens.

This latter claim may have been true when people had CRT monitors on their desks. But it's less valid with today's newer monitors and laptops.

Staring at your phone can also cause vision problems. People frequently report blurred vision, eye strain, dizziness, dry eyes, and headaches after staring at their phones for long periods.

If you suspect your eyes are going bad, you may be right. It's worth considering whether your tech addiction is at least partly to blame.

#9 - Lack Of Impulse Control

Impulse control is the ability to ignore urges. It affects how we interact with everyone and everything around us.

For example, suppose you're standing in line behind two people who are discussing politics. One of them says something that upsets you, prompting an urge to interject. That's natural. It's human nature. But because the two people are strangers, you (hopefully) maintain self-control and ignore the urge.

People who lack impulse control often get themselves into trouble. An example is the driver who experiences road rage when someone cuts in front of him. Or the employee who suddenly quits her job because her boss says something that upsets her.

It shouldn't be a surprise that technology addicts routinely experience lapses in impulse control. They reach for their phones whenever they hear a beep or feel a vibration. They compulsively check their email. They spend hours on Facebook, desperately looking for something to end their boredom.

Their willingness to sacrifice their social lives, long-term health, and happiness to feed their addiction is, by itself, compelling evidence of a lack of self-control.

#10 - Frequent Procrastination

Everyone procrastinates. Again, it's human nature. We prefer to put off tasks we consider to be unpleasant, and instead focus on things that interest us.

For example, at home, we decide to watch television instead of doing housework. At the office, we visit Facebook and Twitter rather than working on the report we're supposed to finish by the end of the day.

Our computers, laptops, and mobile devices are enablers of the procrastination habit. They make it easy to put things off. With a few clicks, we can watch videos, play games, or connect with acquaintances on social media. We can read the latest news headlines, visit our favorite blogs, and text our friends about getting together that evening.

And because we carry our phones everywhere with us, we can do these things on a whim.

Technology puts a myriad of distractions at our fingertips. It's no wonder so many tech addicts have a tendency to procrastinate. Our gadgets provide us with countless opportunities to be blissfully unproductive.

#11 - Increased Irritability

Technology addicts are fine as long as they can surf the internet, play their favorite video games, and watch YouTube videos uninterrupted. But take away their phones and gaming consoles and you'll see a dramatic change in their demeanor. Many become irritable. Some become physically aggressive.

This reaction is common among all addicts, whether they're addicted to sugar or illicit narcotics. Remove the object of their obsession and you'll witness a huge shift in their temperament.

Think back to the last time you were prevented from using your phone, surfing the internet, or playing video games (if you're a gamer). Did you feel agitated? Did you become irritable, perhaps with the person responsible for diverting your attention away from your devices? If so, you may be a tech junkie.

#12 - Impaired Sense Of Time

Losing track of time is one of the most common side effects of technology addiction.

We've all done it. Think about the last time you researched something online that interested you. You may have spent hours reading about the topic without realizing how much time had passed.

The addict experiences this loss of time on a regular basis. Because she spends so much time in isolation staring at her phone, tablet, or computer screen, she often has no idea what time it is. Tech addicts who work in closed spaces devoid of windows or natural light may even neglect meals since their circadian rhythms are thrown off.

If you regularly look up from your gadgets and are surprised by the time, you may have an obsession. If that's the case, you're an ideal candidate for a digital detox.

. . .

IN THE NEXT CHAPTER, we're going to take a look at factors that make you more vulnerable to technology addiction. Before you can control the obsession, it's important to recognize the psychological, physiological, and environmental elements that may be encouraging it.

10 FACTORS THAT INCREASE YOUR SUSCEPTIBILITY TO ADDICTION

~

Technology addiction doesn't happen in a vacuum. There are always identifiable factors that contribute to its development and persistence.

As you'll see in a moment, some of those factors are psychological. They make you predisposed to addiction. That doesn't mean you're helpless. You're not at their mercy. But it *does* mean you'll need to take deliberate, concerted action to break ingrained habits.

Other factors are physiological. The bad news is that there isn't much you can do about your genetics and inherited traits. You're stuck with them. But again, you're not helpless. In upcoming chapters, I'll show you how to control these vulnerabilities and minimize their effects.

Still other factors are environmental. These are things

you *do* control. You can do something about them, and thereby make yourself less susceptible to them.

Following are 10 characteristics that cause, or contribute to, compulsive behavior related to technology. We'll cover them quickly. The goal of this chapter isn't to resolve each factor. Rather, I want to highlight them so you'll be aware of the forces working against you. This will help us focus on the areas that present the most formidable obstacles to completing a digital detox.

#1 - Gender

Studies show that males are more prone to internet addiction and video game addiction while females are more susceptible to phone addiction. Reasons vary and involve issues related to self-esteem, moodiness, inclination to network, and tendency toward social isolation.

While there isn't much you can do about your gender, you *can* take steps to curtail the traits that make you susceptible to certain types of addiction.

For example, if you spend a considerable amount of time alone (social isolation), make weekly lunch plans with a friend. Or join a weekend softball league. Or volunteer at a local charity.

#2 - Mental Health

Mental illness has been linked to many types of addiction. According to researchers, individuals suffering from depression, anxiety, bipolar disorder, and schizophrenia are more likely to have an obsessive personality than individuals not suffering from these conditions.

Such issues warrant seeking the advice and guidance of a mental health professional. I have zero training in that area, so I'm incapable of providing help.

If you suffer from mental illness, I strongly encourage you to seek the appropriate treatment. That's the first step toward controlling the obsessive personality that is making you susceptible to technology addiction.

#3 - Impulsiveness

Do you act without considering the potential consequences of your actions? If so, you have low impulse control, a characteristic we discussed earlier. Importantly, it might be contributing to your internet addiction or phone addiction.

Studies show that impulsiveness is associated with an addictive personality. You're more likely to develop a dependency on your phone, the internet, and social media if you lack the ability to control your impulses.

Part of the problem is that these tools are so accessible. For example, most of us carry our phones with us at all times. We can use them to go online, read the latest news, play games, and visit Facebook the instant we think of doing so.

It's like a heroine addict carrying a vial of heroine in his pocket. He has immediate access to it. Consequently, he's guaranteed to relapse.

The key to controlling your addiction to technology is to curb your impulsiveness. That entails training your brain to self-regulate and ignore urges. A digital detox will prove valuable in that regard.

#4 - Negative Self-Image

The way you see yourself can play a role in whether you develop a technology dependency. If you suffer from low self-esteem or have a negative self-image, you'll be more likely to isolate yourself from others. The isolation will cause you to further question your self-worth. You'll eventually look to your phone and the internet to escape the negative thoughts and seek validation.

This is a common path for those with low self-esteem. The worst part is that it feeds upon itself. A negative self-image causes the individual to seek solace in technology, which reinforces feelings of low self-worth.

Spending time away from your phone and other devices - that is, going through a digital detox - won't eliminate feelings of low self-esteem. It's not a cure-all. But it *will* remove one of the factors contributing to - and even encouraging - such negative feelings.

That's an important step toward changing the way you see yourself.

#5 - Loneliness

Social interactions are essential to our happiness. The relationships we share with our friends and loved ones bring us joy and give us a deep sense of satisfaction.

Loneliness has the opposite effect. When we're lonely, we feel empty inside. We feel separated and disconnected from others. And that can lead to compulsive behavior when it comes to our favorite gadgets.

Persistent loneliness is often a precursor to depression, a common contributing factor in the onset of addiction. Moreover, the absence of a strong social support network increases the likelihood of repeated relapse following recovery.

It's worth noting that being alone is not the same as feeling lonely. Many of us thrive when we're alone, enjoying a high level of happiness and satisfaction. The problem occurs when alone time makes us feel disconnected from others. The absence of social connections opens the door to loneliness and its attendant side effects.

#6 - Peer Pressure

The people with whom we spend the most time are the ones we tend to model. Their thought patterns, decisions, behaviors, and even proclivities influence our own.

This is doubly true for those who have an obsessive personality. They sometimes lose their identities as they adopt the traits of others. It's also true for those who have a negative self-image. They model others' behaviors as a way to gain acceptance and validation.

How does this dynamic relate to technology addiction? If the person being modeled is a "high-frequency" tech user, the person doing the modeling is likely to become the same. Conforming to the model's actions and behaviors means spending more time on the phone, internet, and social media. This is dangerous for an individual who has a compulsive, addictive personality.

#7 - Ease Of Access

Our phones give us instant, continuous access to the internet. That's useful when we need important information, such as driving directions or data we've stored in the cloud.

But it also presents a danger. The ease with which we can go online exposes us to constant distraction. It allows us to feed our obsession for social media, news headlines, cat videos, and video games. It encourages us to randomly surf websites without direction or purpose.

Each time we give in to the temptation to feed our addiction, we reinforce that compulsive behavioral pattern. We thus become more likely to do the same thing over and over again.

The good news is that this is an environmental factor. You have control over it. You can choose to limit your access to your phone, tablet, and other devices, and thereby reduce your dependency on them.

#8 - Isolationism

The more time you spend alone, the more inclined you'll be to seek entertainment, engagement, or validation online. To that end, social isolation can eventually lead to a growing obsession with technology.

Consider that face-to-face interaction triggers the release of dopamine. That's the reason our relationships with friends and loved ones are so gratifying.

In the absence of such relationships, the brain looks for non-social stimuli to trigger the same effect. Time spent online takes the place of face-to-face interaction and replaces the resultant social bonding. Texting, visiting Facebook, watching videos, and even randomly surfing the internet prompts the release of dopamine, flooding the brain's reward center. This is problematic because it creates a dependency, a pivotal step toward developing an addiction.

#9 - Moodiness

A 2015 study in the journal Personality and Individual Differences found that emotional instability was linked to cell phone addiction.

This makes sense. A person whose moods change on a dime is likely to be more inclined to text, tweet, and visit Facebook for quick gratification and mood repair. If the behavior becomes a habit, it's easy to see how it might set the stage for dependency and the onset of addiction.

If you're frequently moody or irritable, it's worth investigating strategies you can apply to become more positive and content. If you feel engaged, optimistic, and hopeful, you'll be less inclined to rely on your phone and the internet for emotional appeasement.

#10 - Addictive Personality

Many people are predisposed to addiction. They're more prone than others to form dependencies. When they encounter conflict or stress, they look to the objects of their addictions for comfort.

For example, an individual who struggles with alcohol might decide to have a beer after a stressful day at the office. A cocaine addict might decide to "do a line" following a heated confrontation with her spouse.

Technology is a perfect enabler for those who have an addictive personality. As previously noted, feeding the addiction, whether by texting, watching videos, or visiting Facebook, provides instant gratification by triggering the release of dopamine. Meanwhile, it poses no immediate nor obvious consequences.

Of course, as is the case with any type of behavior, good or bad, repeated application establishes and reinforces a habit. This can become a thorny situation for the tech junkie, who relies on his gadgets for stress relief.

If you have an addictive personality and have difficulty pulling yourself away from your phone and other devices, it's time to go on a digital detox.

. . .

LET's do a quick recap before we move forward.

You now know the signs of technology addiction. You're also aware of the major factors that contribute to it. In the next chapter, we'll take a close look at the *reasons* for your tech obsession.

Armed with this insight, we'll cover an effective strategy to help you break the cycle and reclaim your time and independence.

THE REASONS YOU'RE ADDICTED TO TECHNOLOGY

~

We use technology to increase our productivity, gain knowledge and insight, and make personal connections on a scale that would be impossible without it. Our phones and other devices enrich our lives in myriad ways.

The problem is, continued use makes us increasingly dependent on them. The more we use our phones, tablets, and laptops for non-work activities - for example, surfing Facebook and Pinterest or playing games - the greater becomes our dependency on them.

That's the road to addiction.

In the previous chapter, we talked about a variety of factors that make you more susceptible to developing an obsession with technology. This chapter will explore the reasons you may be hooked on it. Once you recognize the root causes of your addiction, you'll find it easier to overcome it via a digital detox.

Information Overload

Thirty years ago, information came at us in the form of a stream. We had newspapers, magazines, and a handful of television programs. We visited the library if we needed to research topics.

It was manageable.

Today, information comes at us in the form of a flood. A ridiculous amount arrives through our phones, computers, and hundreds of cable TV channels every day. We increase the load by setting up Google Alerts, subscribing to email newsletters, bookmarking dozens of websites and blogs, and spending hours on social media.

And of course, our Kindles are filled with hundreds, even thousands, of books yet to be read.

We're drowning in information. We're being overwhelmed by a continuous torrent of content.

This keeps us addicted to technology. We struggle to keep our heads above water while at the same time craving more information and entertainment.

Dopamine Rush

The release of dopamine in our brains is the one constant behind all addictions. Studies have shown that playing video games triggers its release. Spending time on Facebook, reading text messages, and searching Google does, too. You experience the same effect when you inhale nicotine, drink caffeine, and take hits of cocaine (or so studies show).

This little neurotransmitter is one of the biggest reasons we become addicted to our drugs of choice, whether we're talking about illicit narcotics or our phones.

Dopamine fills us with a sensation of pleasure. That's a hard thing to turn your back on. Once you experience the sensation, you want to experience it again and again.

Our phones make doing so simple. We receive a text from a friend and feel a small rush. We notice new emails in our inboxes and feel a small rush. We see new activity on Facebook and Twitter, and feel a small rush.

That's dopamine.

It's one of the reasons you may be addicted to your phone, video games, news headlines, and social media. Just as surely as the cocaine addict is always on the lookout for his next hit, the technology addict is always looking for her next fix - whatever will trigger the small dopamine rush her brain craves will suffice.

Anonymity

On blogs, forums, and social media, anonymity equates to safety. We can interact with people, argue with them, ridicule them, and express unpopular ideas, comforted by the thought that no one truly knows who we are.

They don't know our names. They don't know where we live.

It's unsurprising people are willing to say things online they would never say in the physical presence of others. Their anonymity protects them.

Being anonymous also encourages voyeurism. We go online to watch others - some call it stalking - and sate our curiosity about them. What are they doing? Who are they talking to? What are they saying, and about whom are they saying it?

The activity, a combination of people-watching and eavesdropping, can easily become addictive. For the person who already struggles with an obsession with technology, it's likely to feed her habit and reinforce her dependency.

Fear Of Missing Out

The fear of missing out is a powerful motivator. In fact, many companies use it as a sales technique. For example, you've probably seen advertisements claiming that only a limited number of a certain product are available at the advertised price. It's an effective marketing tactic because it provokes our fear of missing out on a rewarding experience (i.e. getting a great deal).

Think about that in the context of reading texts, listening to voice-mails, checking email, visiting Facebook, and reading the latest news headlines. Our compulsion to do these things is driven by the same apprehension. We don't want to miss something.

You've no doubt seen people read - or worse, respond to - texts while driving. You've seen people check their email while sitting in restaurants and movie theaters. They're not addressing emergencies. They're just fearful of missing out.

As I mentioned earlier, the more often we behave in a specific manner - for example, reaching for our phones to read texts the instant they arrive - the more likely the behavior will become a habit. Repetition reinforces the behavioral pattern. Once the behavior becomes a habit, repeated application can turn it into a compulsion. And that's one step away from it becoming a full-blown addiction.

Gateway Of Good Intentions

No one plans to become addicted to their smartphones, tablets, and other devices. We use technology with the intention to make our lives better.

For example, we use spreadsheets at our workplaces to meet our job responsibilities. We set up Google Alerts to save us time on research. We go online to book hotels and airline flights for family vacations.

In other words, technology itself isn't the problem. On the contrary, our phones, computers, and the internet make us more productive and effective. They improve our lives in countless ways.

But counterintuitively, our good intentions can set the stage for the onset of addiction. Individuals who routinely use technology to save time, get things done, or do their jobs may gradually develop an obsession.

Those who have an obsessive personality or are easily distracted are the most vulnerable. The more they use their phones and other tools, the more they reinforce their compulsive behaviors.

For example, checking email becomes less about keeping on top of work-related projects and more about the excitement (and dopamine rush) triggered by the activity.

Societal Expectations

Thirty years ago, people looked curiously - even suspiciously - at those who carried cellphones. The devices were an oddity. And if you owned a laptop, you were considered among the high-tech elite.

Things have changed.

Technology has become a large part of our daily experience. Whether we're at the office, relaxing at home, or on vacation, we're expected to carry our phones and other devices with us. An undercurrent of societal pressure makes us feel naked without them.

This dynamic has made it nearly impossible to function without our gadgets. That's problematic for those who struggle with a phone addiction, internet addiction, or compulsive behavior related to anything involving technology.

We feel pressured to respond to texts and emails the

moment we receive them. We feel pressured to answer our phones when people call us. We feel pressured to keep up on whatever is happening on social media.

Each time we act on our impulses, we reinforce our compulsive behaviors and feed our addiction.

The Tech Industry Encourages Compulsive Behavior

I've likened technology addiction to drug addiction. They share a number of similarities. As noted, engaging in either activity triggers the release of dopamine, a key part of the brain's reward system. That creates and sustains the addiction.

But an addiction to technology is different than an addiction to drugs in at least one important way. And it's one the tech industry relies on to encourage compulsive behavior in consumers.

The fancy term for it is "intermittent reinforcement." It's the practice of using sporadic rewards to reinforce a particular behavior. The term springs from research conducted by psychologist and behaviorist B.F. Skinner in the 1950s.

Here's how it works in the context of checking Facebook. Each time you visit Facebook, you hope to discover something new and interesting. Maybe a friend has posted new photos of her dog. Perhaps your sibling has posted a funny video or a link to an entertaining article.

There's no guarantee you'll find new content that interests you every time you log onto Facebook. That is, there's no guarantee that visiting the site will produce a reward. But if it does so, you'll experience the familiar dopamine rush with its attendant positive feelings.

That's "intermittent reinforcement" at work. According to Skinner, it ensures you'll keep coming back.

Technology companies know this. They use "intermittent reinforcement" as a tactic to encourage your compulsiveness. That includes visiting social media sites, checking your email, reading texts, searching Google, and watching YouTube videos. Each of these platforms are designed to keep you engaged with rewards delivered at irregular intervals.

The same forces are at play when you sit down in front of a slot machine. You don't win with every pull of the lever. You win intermittently. The sporadic rewards keep you in your seat, hoping the replicate the experience.

No Plan To Control The Obsession

The seeds of addiction are planted early. The New York Times reported in 2010 that kids between 8 and 18 years of age spend seven and a half hours a day consuming various forms of media. How? With their phones, tablets, computers, and other devices.

That was several years ago. Technology is playing an even greater role in our lives today. Indeed, CNN reported in late 2015 that teens now spend 9 hours a day watching videos, listening to music, and playing video games. Some of them visit sites like Facebook and Instagram more than 100 times each day.

As one expert noted, "the sheer volume of media technology that kids are exposed to on a daily basis is mind-boggling."

Adults have even more exposure. According to a 2014 Nielsen report, U.S. adults spend 11 hours a day consuming media.

Here's the troubling part: there's rarely a plan, or even an intent, to control this obsession. Parents seldom warn their kids that constant use of phones and other gadgets will reinforce compulsive behaviors and eventually lead to addiction. Indeed, few of us adults take steps to short-circuit that process in our own lives.

Instead, we surrender to the siren call of technology. Rather than limiting our exposure, we surround ourselves with gadgets that continuously feed our obsession.

EVERY ADDICTION, whether to alcohol, drugs, or gambling, carries consequences. An addiction to technology is no exception. In the next chapter, we'll take a quick look at how your tech addiction is negatively impacting your life.

HOW TECHNOLOGY ADDICTION NEGATIVELY IMPACTS YOUR LIFE

~

Most addicts realize their addictions will harm them in the long run. But in the short run, the objects of their obsessions, from drugs and shopping to gambling and chocolate, bring them such a rush that it's difficult to set them aside. Moreover, the immediate consequences of feeding the addiction seem minimal given the immediate payoff.

Technology addiction, like all types of addictions, poses numerous potential consequences. The consequences might seem relatively harmless in the short run, but will introduce major problems in the long run.

Below, we'll address these problems in detail. That way, you'll be aware of the true cost of your obsession when it comes to using your phone, the internet, and other tech-related tools.

Sleeping Problems

In 2014, *Sleep Review: The Journal For Sleep Specialists* reported that cell phone addicts are among the most sleep-deprived individuals. Research has pointed to several possible factors.

First, addicts tend to take their phones to bed with them. This disrupts their circadian systems, or body clocks. Their bodies are consequently less able to regulate when they should sleep and awaken.

Second, the fear of missing out keeps people glued to their phones and laptops long after they should have gone to bed. Because most people need to get up early in the morning, the habit cuts into their sleep.

Third, the type of light emitted by our gadgets' screens - called "blue light" - is believed to play a role. Experts claim blue light tells our brains it's not yet time to go to sleep. Staring at our phones and tablets before going to bed is bound to keep us awake.

Insufficient sleep produces swift, negative side effects. If you're not getting enough sleep, you'll be irritable, more prone to making mistakes, and less able to concentrate. Worse, over time, you'll become more vulnerable to a range of serious health issues.

Increased Restlessness

Have you ever found it difficult to relax because you're waiting to receive an important phone call or email? Have you had difficulty focusing on your work because you're waiting for a friend to respond to your text?

You feel restless.

Restlessness can manifest in a variety of ways. For example, you may be unable to sit still. You might suffer a low level of anxiety. You may be so preoccupied with the object of your attention - e.g. a forthcoming email or text - that you're powerless to focus on anything else, no matter the priority.

Compulsive use of your phone, the internet, and other tech-related tools leads to increased restlessness. Technology allows us to live our lives at a faster pace than ever before. We can obtain information faster. We can connect with, and respond to, friends and family members faster. We can feed our insatiable appetite for new content (videos, blogs, social media posts, etc.) with the click of a button.

This can be an advantage in certain circumstances. For example, if we're trying to meet a tight deadline, and need information to move a project forward, having the ability to obtain that information quickly is valuable.

But this faculty can just as easily become a liability. With every imaginable form of information and every personal connection literally at our fingertips, it's easy to form unrealistic expectations. We expect instant gratification.

When those expectations prove to be false - for example, we're forced to wait for a friend or client to respond to an email - we become restless.

Life is full of such inconveniences. They require patience. Being in a state of constant restlessness is no way to live your life.

Anxiety And Stress

Are you feeling stressed, but unable to identify the reasons? Do you feel anxious despite meeting your personal goals and work-related responsibilities?

It's not a mystery if you spend a considerable amount of time on your phone or the internet. Anxiety and stress are common companions of technology addiction. Studies show that overexposure to the internet can influence your emotional state. It can lead to loneliness and depression, both of which increase your stress levels.

The anxiety and stress occur due to several reasons. Technology addiction disrupts the relationships you share with friends and family members; it causes you to neglect your job responsibilities; it leads to social isolation; and it can eventually open the door to financial problems.

We also noted earlier that an obsession with your phone or the internet can negatively affect your sleep. That too can increase your stress levels. Lack of sleep, especially when sustained over an extended period, triggers the release of cortisol, a stress hormone.

The more stress and anxiety you feel, the less effective you'll be in anything you attempt. You'll be less present when spending time with your family. You'll be less able to concentrate at work. You'll be more irritable and struggle more with distractions. Worse, you'll expose yourself to a long list of health issues, including diabetes, heart disease, high blood pressure, as well as problems with digestion, memory, and sexual function.

A digital detox will help to eliminate the stress you're feeling so you can regain control of your life.

Inability To Focus

We take the ability to focus for granted. We assume we can do it whenever we need to. The reality is that our compulsive behaviors toward technology erode that ability. The more time we spend with our phones and other gadgets, the less we're able to concentrate on whatever task we're trying to complete.

Our thoughts become fragmented. Our attention is quickly lost to distractions. We start to multitask, which further taxes our unfocused brains. Every piece of outside stimuli attracts our notice.

How does this loss of focus impact you? There are several side effects. You become forgetful and miss appointments and deadlines. You become a less engaged - and less engaging - conversationalist. Your situational awareness suffers. You become more impulsive and prone to outbursts. You require more time than necessary to complete projects and activities.

In short, an inability to focus stemming from an addiction to technology can have long-ranging negative impacts on your life.

There are numerous strategies for improving your focus. One is to unplug. If you're struggling with an internet addiction, a phone addiction, or any other tech-related obsession, you need a digital detox.

Inability To Retain Information

Short-term memory loss is often seen in those who struggle with internet addiction. Scientists aren't certain of the reasons, but suspect they might be related to changes in the brain's structure.

In 2011, researchers in China studied the effects of internet addiction on brain gray matter. They found evidence that overexposure to the internet resulted in "multiple structural changes in the brain." They went on to note that "structural abnormalities in the internal capsule could consequently interfere with the cognitive function and impair executive and memory functions." (Their findings were published in the journal PLoS ONE.)

It's tempting to dismiss signs of memory loss as inconsequential. For example, we assign little significance to our inability to remember where we put our car keys. But unless the forgetful individual takes steps to prevent further memory erosion, "inconsequential" signs can introduce troubling circumstances down the road. Loss of short-term memory may gradually influence his emotions, motor function, reflexes, and ability to communicate.

Scientists continue to study how internet addiction affects the brain's numerous processes, including short-term memory storage. If you struggle to remember things and have a compulsion to go online, now is a good time to curb the habit.

Greater Susceptibility To Distractions

By now, it shouldn't surprise you that an addiction to your phone or the internet will make you more vulnerable to distractions. The worst part is that the effect is insidious, and thus often escapes notice.

It happens to everyone. In 2015, Pew Research released its findings concerning smartphone use in the U.S. Nearly 60% of survey respondents reported feeling regularly "distracted" as a direct result of their phones.

It's reasonable to assume the actual figure is higher. After all, many people are disinclined to admit any type of shortcoming. Moreover, some phone addicts may be unaware of the problem.

Our phones can also present the seductive mirage of increased productivity. Many of the survey respondents noted that using their phones made them *feel* as if they were more productive despite increased distractions.

Distractions are the enemy of productivity. Interruptions destroy your momentum, increasing the time it takes you to get something done. Whether the interruption comes in the form of a coworker with a question or an urge to check Facebook, it causes your brain to stumble. Each time this happens, your brain needs up to 20 minutes to get itself back on track.

With that in mind, imagine the daily experience of a technology addict. She's constantly on her phone or surfing the internet. Every few minutes, she checks for new emails and texts, plays games, logs into social media, reads the latest news, or visits her favorite online forums.

Can you imagine how difficult it would be for her to focus on her work or the person she's with? The constant distractions would make doing so impossible.

Being distracted by your phone also poses other issues. For example, it impairs interpersonal bonding. Many people consider it rude to acknowledge texts, emails, and phone calls in their company. Additionally, using your phone while driving can have catastrophic results.

If you're addicted to your phone, you probably don't consider the constant distractions to be a problem. At least, not yet. That perspective will change once you go through a digital detox. You'll notice a dramatic difference in your level of awareness in the absence of your gadgets.

Reduced Productivity

The more often you're distracted, the less productive you'll be. Given that a phone obsession or internet addiction opens the door to an endless string of distractions, your productivity is sure to decline.

Many people mistakenly assume their gadgets make them more productive. The perception is understandable. After all, they can check their email, manage their schedules, and respond to questions instantly via text.

But the perception is oftentimes an illusion. More often than not, our phones and other devices serve as obstacles to our attempts to get things done.

The economist Robert Solow addressed this notion in 1987 when he noted *"You see the computer age everywhere but in the productivity statistics."* Although the advent of smartphones, tablets, and wi-fi internet access was still several years away, Solow's comment was insightful. Even prescient.

The truth is, many of us waste a considerable amount of time online each day. We're constantly distracted, with our attention pulled in a thousand directions. The result is that we're regularly unable to get as much done as we hope.

Think about how you currently use the internet. You might spend an hour each day reading and replying to emails. You may spend an hour each morning reading news headlines and watching YouTube videos. Perhaps you spend a few hours in the afternoons playing online games, surfing social media sites, and reading your favorite blogs. Maybe your evenings are spent shopping or gambling online.

These activities can take up a considerable portion of your day. That leaves you with significantly less time to complete important tasks and projects.

Over the past 10 years, there has been a surge of productivity apps you can download to your phone and browser. These apps are seductive because they hold the promise of increased productivity. But the developers rarely mention that becoming more productive is more a matter of developing good habits than simply downloading the latest apps.

If you're currently struggling to finish projects on time, consider how much time you spend on the internet and your phone. It's possible you've developed an obsession. If that's the case, your compulsive behaviors are almost certainly hampering your productivity.

Strained Relationships

Technology addiction can take a dramatic toll on the relationships you share with the people in your life. Consider the phone addict who checks her phone every couple of minutes. She's unable to carry on a meaningful conversation due to the constant interruptions. Understandably, the person she's with is likely to think of her behavior as rude.

Consider the internet addict who can't pull himself away from his computer. He's obsessed to the point that he'd rather stay online than retire to bed with his spouse.

Technology addiction impairs relationships in numerous ways. Some are less than obvious.

For example, it hampers the intimacy shared between spouses. An endless string of text messages, email notifications, and event reminders creates an ever-present distraction that makes communication and intimacy difficult, if not impossible.

It causes the person spending time with the addict to feel devalued in his eyes. As the addict continues to check his email and texts, he neglects his companion, who eventually realizes that he or she is a lower priority than the addict's phone.

Technology addiction also causes us to lose our empathy for others. We become less capable of understanding what they're experiencing. We lose our ability to sympathize with them. It follows that we become less able to grasp how our compulsive behaviors affect them. We develop a social blind spot. The blind spot makes it difficult for us to connect - or maintain connections - with others, including the people who are most important to us.

Erosion Of Interpersonal Skills

We assume technology improves our communication. And in many ways, it does.

For example, we're able to reach out to people whenever the mood strikes or need arises. We're able to connect with them whenever we want since most people carry their phones with them. If we need something from someone and don't feel like having a conversation, we can send a concise email or text.

But there's a dark side to this capability.

The more we engage online, the more our interpersonal skills deteriorate. The more we interact with our friends and loved ones through texts, emails, and social media, the more we dilute the real-world connections we share with them.

A lack of interpersonal skills poses social consequences.

First, it impairs communication. Second, it makes one less inclined to listen to others, a vital tool for making personal connections. Consequently, an individual with poor interpersonal skills is likely to struggle in group settings and circumstances involving teams.

An erosion of interpersonal skills is one of the many ways a technology addiction erects a social barrier around the addict. The effect is significant. The stronger the barrier, the more isolated the addict will feel, setting the stage for anxiety, loneliness, and depression.

WHAT IS A DIGITAL DETOX (AND HOW WILL IT HELP YOU)?

∾

There are a few ways to treat technology addiction. One way is to seek counseling from a therapist. This can be done in a private or group setting.

Another option is rehab. This approach works in the same was as drug or alcohol rehab. You visit a treatment center for a specific period of time - for example, two weeks - during which your access to technology is restricted. Some clinics offer outpatient programs, but as with outpatient drug and alcohol rehab, there's a high rate of relapse.

The third approach is self-treatment. Rather than visiting a therapist or signing yourself into a rehab facility, you take the reins in breaking your addiction. You're in control.

Here, we're talking about doing a personal digital detox.

The advantages of self-treatment are twofold. First, it's less expensive than therapy or enrollment into a rehab facility. Second, it poses less disruption to your current lifestyle.

If you're like me, those two advantages seal the deal.

Below, I'll explain what a digital detox is, and present the reasons to do one. If you're addicted to text messaging, video games, social media, news headlines, or YouTube, you're about to discover one of the best remedies.

Digital Detox Explained

A digital detox involves stepping away from all of your gadgets. This includes your phone, tablet, and laptop.

There are a number of challenges to traveling this path. For example, how do you unplug completely if your job requires you to use a computer? Additionally, how can you keep in touch with people without texts, emails, and phone calls?

And what about withdrawal symptoms? A digital detox is like a drug detox. Your brain craves the dopamine rush that results from your compulsive behavior. Once you cut off your access to technology, you'll feel the pangs of withdrawal.

I'll address these and other challenges in the following chapters. The important thing to remember is that breaking your addiction requires the removal of your phone and other tech-related tools. Weaning yourself is not enough if you hope to control your dependency. You need to sever the connection, at least temporarily.

Think of it this way: if you were addicted to cocaine, you wouldn't try to break your addiction by gradually reducing the number of lines you do each night. That strategy is guaranteed to fail. Instead, you'd check yourself into a clinic and undergo a complete detox, one with no access to cocaine.

That's how you need to approach your addiction to technology. If you want to break the habit, stop feeling overwhelmed, and regain control of your focus and productivity, you need to do a complete digital detox.

You'll probably need motivation. I strongly encourage you to write down the many ways your life will improve after you break your technology addiction. Review the list whenever you experience signs of withdrawal.

I'll help you get started. The next chapter will highlight several reasons to do a digital detox.

15 WAYS YOUR LIFE WILL IMPROVE AFTER A DIGITAL DETOX

~

I won't lie to you.

Going through a digital detox can be unpleasant - at least, in the beginning. You'll crave your phone so you can check for new texts. You'll covet your laptop so you retrieve your email. You'll lust after your tablet so you can read the latest news headlines and log onto Facebook.

And you'll grow increasingly restless and agitated the longer you're prevented from doing these things.

These are symptoms of withdrawal.

You need to have a good reason to cut yourself off from your phone and other gadgets. That reason will help you to resist the temptation to reach for them when the cravings start. You need to know that the effort will improve your life in some measurable way. Otherwise, there's no reason to move forward.

Below, I'll point out 15 ways your life will improve after you go through a digital detox. Some of them may seem inconsequential to you. Others will be of great importance. Your circumstances are unique to your life. So note the improvements that matter most to you.

#1 - You'll Break The Cycle Of Addiction

The cycle of addiction is driven by the reward center of the brain. That's the reason it's so tough to break. As we noted earlier, dopamine floods the limbic system of the brain whenever the addict feeds his addiction. This effect hijacks the brain, producing intense cravings. The cravings grow in intensity until the addictive substance - for our purposes, anything related to technology - is used.

When you go on a digital detox, you break this cycle. You interrupt the behavioral patterns that have, until now, reinforced your addiction. The detox gives you an opportunity to "reset" your brain's reward system.

#2 - You'll Form Deeper Connections

As we noted earlier, checking your phone every few minutes hampers your ability to connect with those around you. So does spending all of your free time online rather than interacting with people face to face.

When you set aside your phone and other devices, you'll develop a heightened sense of empathy. You'll be better able to read people's faces and body language and recognize the emotions they're experiencing. You'll also enjoy uninterrupted time to have meaningful conversations with them.

The result?

You'll feel more connected to others. You'll be able to give people the attention they deserve. In return, you'll receive more attention yourself. Forming such connections can be a deeply rewarding experience.

#3 - You'll Reclaim Valuable Time

How many times do you check your phone each day for new messages? How often do you stop what you're doing to visit YouTube, Facebook, or other websites unrelated to the whatever you're working on?

You're probably wasting at least a few hours each day.

I know this firsthand because I used to be addicted to news. I'd check news sites like CNN.com every few minutes. Although I was up on current events, my productivity was terrible. I eventually went on a news "fast" to reclaim the time.

When you go through a digital detox, you'll enjoy the same benefit. You'll recoup the time you once spent checking your phone, visiting Facebook, Twitter, and Instagram, and watching video after video at YouTube. You'll probably find that you spent more time doing these things than you had imagined.

#4 - You'll Be Able To Think More Clearly

If you're a phone addict, your phone is a constant distraction. Its beeps and buzzes, which notify you of incoming messages and calls, are impossible to ignore.

If you're addicted to social media, you'll feel the unremitting lure of sites like Facebook, Twitter, and Instagram. They're always calling to you, tempting you to drop whatever you're doing to log on.

If you're obsessed with the internet, every moment you sit at your computer will be a struggle. You'll be constantly tempted to set aside work to surf your favorite websites.

These compulsions make it difficult, and even impossible, to think clearly. You're in a state of continuous distraction. You'll also face the temptation to multitask, an activity that research shows is contrary to how the brain works.

A digital detox clears the mental deck. It wipes away the technology-related distractions so you can relax and think clearly.

#5 - You'll Enjoy Improved Focus

In the chapter *Top 12 Side Effects Of Technology Addiction*, I noted that goldfish, on average, have a longer attention span than people. Of course, the fish have an advantage - they lack access to smartphones, social media, video games, and YouTube.

These "tools" are distractions if you're addicted to their use. And a distracted mind is an unfocused mind.

Have you been feeling unfocused lately? Do you have trouble concentrating on your work, and consequently make mistakes? When your spouse or children talk to you, does your mind drift?

If so, you're not alone. In this age of digital media, millions of people struggle with the same challenges.

The quickest, simplest way to regain focus is to get rid of your distractions. It's not enough to lessen their number or reduce your exposure to them. You need to purge them entirely. If you're among the technology-obsessed, that means setting aside your phone, the internet, your gaming console, and any other "tool" you rely on for your fix.

#6 - You'll Be More Productive

Have you ever been in a flow state - it's sometimes called "being in the zone" - while working at your desk? It's a state of mind where you have laser-sharp focus. You're also excited about your work, which increases your energy level. As a result, you're able to get a heap of work done seemingly without effort.

If you're a writer, being in a flow state means writing while the perfect words and passages pour out of you. If you're a student, it means working on a project without any sense of time or mental exertion. If you're a web developer, being in the zone means creating a new design framework with elements that seem to fall right into place.

Working in a flow state leads to greater productivity. It allows your brain to process information quickly and efficiently. At the same time, stress evaporates, allowing you to focus exclusively on the task in front of you.

How do you achieve a flow state? The first step is to rid your environment of distractions. To that end, if you're normally distracted by your phone, Facebook, YouTube, and video games, you need to do a digital detox.

#7 - You'll Sleep Better

It shouldn't surprise you that technology addiction impairs the quality of your sleep. Plenty of research backs up the connection.

For example, a 2015 study published in the Journal of Behavioral Addictions showed that phone addiction is positively correlated with poor sleep. In 2012, the Journal of Sleep Research published findings from a study that indicated a strong link between internet addiction and insomnia. In 2016, a study examining the effects of internet addiction disorder (IAD) on Korean college students found the condition was strongly linked to insomnia and poor sleep quality.

If you find it difficult to set aside your phone or log off the internet, there's a good chance you're addicted. Your addiction is probably causing you sleeping problems, even if you don't realize it.

If you constantly feel tired, monitor whether you're getting enough rest at night. If not, your favorite devices may be to blame. They might be keeping you awake, making you an ideal candidate for a digital detox.

#8 - You'll Avoid Information Overload

As a technology addict, your brain is overloaded with information. Checking your phone for messages, reading your favorite blogs, watching YouTube videos, and retrieving your email takes a toll. If you're doing these things every few minutes, your brain never gets the time it needs to recharge.

That's a serious problem. We drink from a fire hydrant of information online, assuming it will make us more knowledgeable. In reality, we allow ourselves to be bombarded with content that is irrelevant to our goals.

The result?

We're less focused, more stressed, and need extra time to make decisions. Additionally, the deluge of information distracts us, making us more inclined to multitask and more likely to make mistakes.

Being selective of the information you consume is one of the keys to maintaining a high level of productivity. That will become crystal clear once you go through a digital detox. You'll discover that much of the content you normally consume via your favorite devices is unnecessary.

#9 - You'll Experience Less Stress

If you're feeling stressed, the reason may be your use of the internet or your phone. Studies show that constant use of technology can lead to higher levels of anxiety.

For example, in 2014, the International Journal of

Humanities and Social Sciences published research showing a significant relationship between internet addiction and stress in college students. The more advanced the addiction, the greater the level of stress observed by the researchers.

That's unsurprising. After all, think about how you use your phone and the internet. If you're an addict, you probably spend at least 10 hours a day compulsively using these tools. That volume of use exerts stress on your mind and body.

In addition, people who constantly use their phones for social purposes are prone to experience "social stress." Daily interactions with others, whether via text, email, or phone calls, cause anxiety and gradually become undesirable.

As we covered earlier, stress can lead to a variety of secondary problems. It can impair your sleep, affect your digestion, and cause aches and pains. It can wreak havoc with your mood and even set the stage for depression.

Spending time away from your phone and other devices will relieve much of the stress you're experiencing.

#10 - You'll Adjust Others' Expectations Of You

Technology addicts tend to respond quickly to texts, emails, and voicemails. They're compelled to check for new messages every few minutes. When they notice they've received one, they're unable to resist sending an immediate response. Doing so makes them feel productive, which reinforces the habit.

There's a terrible downside to this practice: people become trained to expect an immediate response.

You may already be dealing with this problem. When people send you texts, they expect to hear back from you within moments. When they send you emails, they expect a reply that same hour. And if you don't respond to their voicemails within an hour or two, they're likely to call again to find out what's wrong.

If you struggle with an addiction to technology, these expectations are bound to increase your stress level. They'll also distract you from whatever task or project you're working on since you'll constantly be reacting to external factors.

One of the advantages of going through a digital detox is that you'll reset other people's expectations. Without access to your devices, you won't be able to get back to them in as timely a manner as they've come to anticipate.

That's fine. Emergencies aside, there's rarely a good reason to reply to a message the moment you receive it. Some messages should be left to simmer, giving you time to think of suitable responses. Some warrant no responses at all.

#11 - You'll Develop Better Social Skills

Studies show that constant exposure to digital media has impaired our interpersonal skills.

In 2014, the journal Computers in Human Behavior compared two groups of preteens. One group had constant access to their phones and other devices. The other group had no such access. The authors found that the latter group enjoyed increased social interaction since its members were spending less time with their gadgets. As a result, they were better able to recognize nonverbal, emotional cues, a critical skill in establishing social bonds.

Adults are also impacted by the social consequences of overexposure to digital media. In early 2014, Justine Harmon, senior editor for the magazine *Elle*, wrote a feature titled "How Social Media Is Ruining My Social Skills." She noted that spending time on websites like Instagram had begun to affect how she interacted with people she met face-to-face. In a Huffington Post interview following her feature, she opined that overuse of social media is "the death of an actual civilized conversation."

Online networking, whether done through email, texts, or social media, is a completely different dynamic than interacting face to face. Other online pursuits, such as watching YouTube videos, playing video games, and reading news headlines, offer no social interaction at all. It's no wonder so many people have difficulty starting and holding conversations.

A digital detox will improve your ability to communicate face to face. Without access to your phone, computer, and other devices, you'll be forced to rely on your interpersonal skills. This presents a good opportunity. You'll become better at listening, maintaining eye contact, and reading body language.

#12 - You'll Enjoy Reading Books More

Do websites, blogs, forums, texts, and emails comprise the majority of your daily reading material? They do if you're a bona fide technology addict.

This might seem fine at first. After all, reading of any type informs us. It entertains us. It can even help us to forget about the things that are causing us stress.

But reading online articles and blogs isn't the same as reading a novel. The latter delivers unique benefits. When you read a fiction novel, whether a mystery or high-octane thriller, you stimulate your brain, improve your focus, and better your memory. You also expand your vocabulary and consequently, enhance your ability to communicate.

Studies also show that reading for pleasure, typically by way of a good novel, can help you to relax and even improve the quality of your sleep. Moreover, some types of fiction offer insight into others' mental states, making you more compassionate and empathetic.

How long has it been since you've read a novel from cover to cover? Months? Years? When you set aside your devices (your Kindle is an exception), you'll rediscover the joy of reading. You may even find that you want to continue the habit after you complete your digital detox.

#13 - You'll Have More Time To Exercise

I realize some people read and send texts, check their email, visit social media, retrieve their voicemails, and return calls while they exercise. But they're in the minority. For most, it's an "either-or" proposition. They either spend their time using their devices or they work out.

They rarely do both at the same time.

In most cases, given the choice, tech addicts will always choose technology over exercise. Phone and internet addicts are more inclined to visit Facebook, watch YouTube videos, and play video games than they are to go for a jog. They're more disposed to text their friends than they are to set aside their phones and do pushups and crunches.

And what about those folks who visit the gym with their phones in their hands? Their attention is divided between their workouts and their gadgets. You can tell by watching how they behave. Most just go through the motions of working out while staring at their phones.

A digital detox allows to focus on your physical fitness. Without your phone and the internet holding your attention captive, you'll be better able to concentrate on your workout.

The result? Better health, reduced stress, improved sleep, and a more consistent mood. And that's just the tip of the iceberg.

#14 - You'll Give Your Brain A Much-Needed Rest

Show me a technology addict and I'll show you someone who habitually multitasks. That's a problem. Multitasking, particularly with technology, puts an enormous strain on the brain. Between reading and responding to texts, checking email, logging onto social media, playing video games, and reading the news, in addition to addressing work-related responsibilities, our brains are in constant overdrive.

That's unhealthy. The brain needs to rest. Overstimulation makes it less effective, impairing its performance.

With numerous external factors demanding the addict's attention, she becomes less able to concentrate and more susceptible to distractions. In addition, the continuous barrage of media hampers her ability to store new - and remember old - information. Many tech addicts also struggle with making decisions. They're bombarded with details, and thereby become paralyzed with uncertainty.

Your brain needs to rest. If you've been pushing yourself mentally, it needs to recuperate in order to perform well later. Setting down your phone and other devices will go a long way toward giving your brain much-needed downtime.

#15 - You'll Develop Better Impulse Control

If there's one thing technology addicts struggle with, it's impulse control. Impulse control is what allows us to make rational decisions regarding our limited resources (e.g. time, money, etc.) in light of our needs and desires.

For example, let's say you're meeting your parents for breakfast. Unfortunately, you're addicted to social media.

If you have good impulse control, you'll be able to ignore notifications alerting you of new Facebook posts and meet your parents on time. If you have poor impulse control, you'll check every notification as it arrives, delaying yourself and ultimately making you late for your breakfast date.

There's an entire field of psychology devoted to the mechanisms involved with impulse control. For the purposes of this action guide, it's enough to say that over-exposure to technology is a contributing factor to its decline among millions of people.

Scientific research supports the connection. In 2016, the journal Psychonomic Bulletin & Review published a study showing that heavy phone use was linked to poor impulse control and a diminished willingness to delay gratification.

If you struggle with this issue, rest assured you can learn how to control your impulses. That's one of the benefits of going through a digital detox. You won't have access to the devices that feed your habit.

It's Not Your Fault

If you're struggling with a phone or internet addiction, it's important to realize that it's not your fault.

Our phones, iPads, and laptops are part of our everyday experience. It's nearly impossible to live without them because we're expected to be reachable at all times.

Moreover, if we're not interacting with our peers through digital media, whether via texts, emails, or social media, we feel disconnected. Even isolated.

The reality is that our high-tech gadgets are forming a wall between us and the people that are important to us. The more time we spend online, the higher and thicker the wall becomes. We risk damaging the relationships we share with our friends and loved ones, and consequently hampering our personal growth and happiness.

Fortunately, there's a solution. The remainder of this action guide will take you through the steps of a digital detox. We're going to move quickly so you can plan and execute your own detox as soon as possible. In the next chapter, I'll explain what to expect from the experience.

WHAT TO EXPECT DURING YOUR DIGITAL DETOX

~

As I mentioned in the last chapter, going through a digital detox can be unpleasant. You'll experience symptoms of withdrawal, much like an alcoholic or drug addict enrolled in rehab. Your brain's reward system has grown accustomed to your addiction. Cutting off the source of your addiction is bound to trigger cravings.

If this is your first digital detox, you'll find it helpful to anticipate the symptoms you're likely to experience. I'll cover them below. I'll then describe how you'll feel after you've completed your detox.

Symptoms You'll Experience During Your Digital Detox

Technology gives you an immediate sense of gratification. As we've covered, that's a result of dopamine flooding your brain. Every time you succumb to the compulsion to use your phone and other devices, you reinforce your brain's expectation of future gratification.

That's how a habit becomes an addiction.

When you remove the source of that dopamine release, your brain rebels. It organizes a mutiny, triggering symptoms that pressure you to give in to your dependency.

Following are the symptoms you're likely to face.

Yearnings For Your Phone

This goes beyond simply *wanting* your phone. You'll experience a deep, ravenous hankering for it. You'll feel as if you're unprepared to face the day without it.

At various points throughout the day, you may instinctively reach for your phone only to realize it's not there (it's absence is key to an effective detox). That will cause an immediate sense of disappointment, which will only inflame your longing for it.

Phantom Phone Syndrome

You'll feel as if your phone is vibrating in your pocket even though it's not in your pocket. You'll hear your phone ringing despite the fact that you've turned it off.

These are phantoms. Scientists claim they're a result of your brain misinterpreting sensory input. Psychologists claim the phantom vibrations and ringing are the result of anxiety stemming from deprivation. That is, your brain, deprived of the gratification that occurs when you use your phone, experiences stress. The stress causes it to misinterpret sensory input.

Cravings For Access To The Internet

If you spend the majority of your day and evening online, spending several hours offline is almost certain to trigger signs of withdrawal. You'll become anxious and look for opportunities to obtain access to the internet. You'll feel increasingly agitated and out of sorts if you're unable to do so.

If others are looking at their phones, you'll look over their shoulders to catch a glimpse of their phones' screens. If others are working on their laptops, you'll casually - or not casually, depending on your level of anxiety - look at their screens to see what they're reading.

Your cravings may grow so strong that you feel as if you're unable to function. But that's a mirage. It's your brain doing it's best to pressure you to give in to your addiction.

Irritability & Moodiness

These two symptoms are common in all addicts under-going a detox. Together, they can put the addict, along with the people she loves, on an emotional rollercoaster.

There are a few theories regarding what triggers the irritability and moodiness. Some experts claim the addict engages in her compulsion - for example, overuse of her phone - to escape her emotions. When she is deprived of that activity, her emotions flood back in and overwhelm her. Other experts note that a detox makes the addict's mind more fragile. Combined with the intense cravings stemming from withdrawal, she's less able to handle vexatious circumstances.

You're likely to experience these same symptoms. Their severity will depend on the extent of your technology addiction.

Headaches

Headaches are also common during a digital detox. They usually stem from tension. The anxiety you'll feel during the first stages of your detox can literally make your head hurt.

You'll probably find that the headaches go away soon after you start your detox. That's normal. The more time you spend away from your gadgets and the internet, the more your brain will grow accustomed to the "silence." As it does so, the tension will dissipate.

Drowsiness

When you're using your phone and other devices, it probably seems as if you're fresh and wide awake. In reality, you're stimulating your brain and refusing to let it rest.

This is one of the reasons many technology addicts undergoing a digital detox experience drowsiness. Without their gadgets providing constant stimulation, their brains are finally able to relax.

Another reason for the drowsiness is that tech addicts often have difficulty sleeping. The lack of restful sleep eventually catches up to them. In the absence of their devices, their brains take a much-needed breather.

Inability To Concentrate

During a digital detox, it's common to have difficulty thinking clearly. That's doubly true if this is your first detox. You'll be unaccustomed to the fuzziness in your mind during the early stages.

The inability to focus is due to the manner in which your brain has adjusted to your technology-related compulsions. It has grown used to your addiction. When you cut yourself off from your phone, the internet, video game console, and other devices, your brain is forced to make adjustments. That causes a mental fog.

You might have trouble remembering things and processing new information. Your attention span might wane as your thoughts are pulled in numerous directions.

You might feel slightly disoriented and even have trouble communicating.

THESE SYMPTOMS usually present during the first stages of a digital detox. It's important to remember that they're temporary. Once they dissipate, you'll feel much better.

In the next chapter, I'll describe how you'll feel after you've completed your detox.

HOW YOU'LL FEEL AFTER YOUR DIGITAL DETOX

~

Once you're done with your detox, you'll enjoy several benefits that might have once seemed alien to you. Here's a smattering of what you can look forward to:

You'll Feel More Connected To Others

Without technology dominating your attention, you'll be better able to maintain eye contact, listen more carefully, and enjoy more meaningful conversations. You'll create stronger bonds with your friends and family members. You'll more easily connect with strangers.

You'll find that the surface-level "friendships" you've formed online - for example, on Facebook - pale in comparison. They don't have the same depth or intimacy.

You'll Have A Sense Of Inner Peace

If you live on your phone or computer, you probably feel harried. That's normal among tech addicts. The constant pressure to respond to texts and emails, combined with the fire hydrant of information delivered to you 24 hours a day, causes stress.

You might have attributed the stress to your job, financial situation, or any number of personal obligations. While these items may indeed be stressors in your life, the sense of anxiety you're feeling is exacerbated by your overuse of technology.

When you give up your gadgets, even just temporarily, you'll experience a sense of calmness. Everyone does. You'll be free of the constant agitation that comes with compulsive use of your devices.

You'll Feel Untethered To Others' Expectations

Think about how you respond to texts, emails, and phone calls. Do you reply to messages and return calls immediately? Do you expect to be questioned if you fail to do so within an hour? Are you sometimes asked to explain yourself if you make people wait a few hours, or even a full day, before you respond?

One of the greatest advantages of "unplugging" is that you'll remove the weight of other people's expectations concerning your response time. You'll force them to adjust their expectations.

This may seem unimportant. But consider the freedom it gives you. You'll no longer feel pressured to respond instantly to texts. You'll feel at ease letting noncritical emails sit for 24 hours. You'll feel comfortable returning calls according to *your* schedule rather than the caller's.

You'll Enjoy Renewed Interest In Forgotten Hobbies

Did you used to enjoy a hobby that has since been relegated to the "when I get the time" pile? Perhaps you liked to garden, paint, write, or complete jigsaw puzzles. Maybe you enjoyed hiking, bird watching, or woodworking.

Whatever your past interests, you're no longer doing them. And if you're like most people, you're probably telling yourself that you lack the time.

But it's likely your addiction to technology has played a huge role. Your time, attention, and energy, once devoted to your hobbies, are now devoted to your phone and online activities.

As previously noted, a digital detox severs the connection. It resets your behavioral patterns. Once you've completed your detox, you may find that the hobbies that once held your interest are more compelling than the compulsion to use your devices.

You'll Experience The Joy Of Missing Out

In the chapter *Reasons You're Addicted To Technology*, I mentioned the fear of missing out as a contributing factor in your addiction. It's the phobia that you might miss out on a rewarding experience.

This constant sense of angst spurs many tech addicts to stay "plugged in." Unfortunately, doing so creates a subtle, continuous undercurrent of anxiety in their lives.

After completing a digital detox, you'll experience the opposite effect: the *joy* of missing out. You'll no longer care whether you're informed about the latest current events. You won't care whether you're "in the know" about the best parties in your city. You'll lose interest in constantly checking your phone for new messages just in case your friends are doing something fun without your knowledge.

The joy of missing out allows you to relax and relish your downtime. You'll appreciate quiet and solitude, and even learn to treasure activities you once found boring - for example, reading a good novel.

If these benefits sound appealing to you, realize they're within your grasp. You can improve your relationships and enjoy a sense of inner calmness. You can reset others' expectations of you, reignite your passion for neglected hobbies, and experience the relaxed, stress-free state that accompanies the joy of missing out.

How? By going through a digital detox.

The next several chapters will show you how to prepare for the experience. Each one is short, covering a single step. They're organized in such a manner so you'll be able to find them easily in the future.

PART I

HOW TO PREPARE FOR A DIGITAL DETOX

~

Alexander Graham Bell once said *"Before anything else, prepa-ration is the key to success."*

That's certainly the case when doing a digital detox.

As with beating any addiction or breaking any type of compulsive behavior, severing your dependency on tech-nology requires a plan. You need to set things up properly to support your endeavor.

That's the purpose of the following nine steps. They'll minimize the negative effects of unplugging - for example, your friends' frustrations at being unable to reach you via text messages. They'll also help you to resist the tempta-tions you'll face in the absence of your favorite devices and access to the internet.

We're going to move quickly through these nine preparatory steps. Don't dismiss them for their simplicity. Each one is important. Together, they'll help you to successfully complete your digital detox.

PREP STEP 1: CREATE A DIGITAL DETOX PLAN

~

Ask yourself two questions. First, how long will your digital detox last? Second, when will your schedule best accommodate it?

Many people who struggle with technology addiction are overly aggressive when it comes to deciding the duration of their detoxes. For example, they plan to unplug for an entire week, or even longer.

Their enthusiasm is understandable. They're tired of allowing their gadgets to run their lives and want to finally break the habit. But their expectations are unrealistic.

Unless you're going on vacation, a 7-day technology fast is impractical. After all, if you're like most people, you need access to your computer to do your job. You also need your phone so your boss and coworkers can reach you in the event of a work-related emergency.

Here's an alternative: rather than planning a 7-day digital detox, plan one that will last for just 24 hours. Most people can fit that into their weekends.

That brings us to the second question concerning when your schedule will best accommodate your detox. Assuming you work Monday through Friday, the weekends will be your best option.

Choose one day and mark it on your calendar. That will ensure you remember it. Putting it on your calendar in plain sight will also prompt your brain to prepare for it.

A digital detox is not a one-time affair. Our lives are so immersed in technology that it's useful to do a detox two or three times a year. Stick with a 24-hour detox for your first time. Once you've experienced its benefits, feel free to extend future detoxes to 48 hours, and even 72 hours over a 3-day weekend.

One quick note: feel free to use your Kindle or Nook during your detox. But limit their use to reading books.

If you own a Kindle Fire, set it aside until you complete your detox. Why? Because the Fire is designed to deliver much more than just books. It allows you to watch movies, listen to music, use apps, and play games.

In short, it'll pose too big of a temptation.

PREP STEP 2: REMOVE SOCIAL MEDIA APPS FROM YOUR PHONE

❦

This step may seem like overkill. After all, if you're setting your phone aside, why would you need to get rid of your social media apps?

Two reasons.

First, I know from experience how difficult it is to ignore your phone during a detox. It's like an extra appendage. You feel incomplete without it.

Your first instinct will be to reach for your phone when you experience cravings. In the event that happens, you'll be less tempted to log onto Facebook, Twitter, and Pinterest if those apps have been deleted.

The second reason to delete these apps is that doing so eliminates their respective push notifications. These are the alerts that let you know about updates on Facebook and other social media sites.

You may be thinking *"If my phone is off, why do I need to worry about the notifications? I won't see or hear them."*

Because emergencies happen. In the event of an emergency, you'll have a valid reason to use your phone - for example, to make sure your kids are okay - during your detox. The problem is, when you turn it on, the notifications from your social media apps will display, triggering your fear of missing out. If you're obsessed with Facebook and Twitter, it will take a gargantuan amount of willpower to resist the temptation to check in.

It's best to delete the apps from your phone to avoid the temptation altogether.

There are many apps designed to either block notifications or block access to social media sites for a specific length of time. The problem is that you can override them. And given your obsession with technology, there's a good chance you'll do so.

Imagine being addicted to ice cream and putting a padlock on your freezer. If you have the key, there's a high likelihood you'll remove the padlock when you experience strong cravings. Willpower isn't enough. Thus, it's best to remove the ice cream altogether.

The same principle applies to beating your technology addiction.

Bottom line: remove your social media apps. You can always reinstall them after completing your digital detox. Or if you're like me, you'll find that life is more enjoyable without them.

PREP STEP 3: CREATE AN OUT-OF-OFFICE EMAIL MESSAGE

~

I f you're normally reachable via email, your friends, family members, and coworkers are going to wonder where you are during your detox. They'll notice you're not responding to their emails as quickly as they've come to expect from you.

The solution is to create an out-of-office message in your email program. This message will be sent out automatically and immediately each time you receive an email. The sender will receive the message informing him or her that you'll be away from your computer until a specific date.

What should you include in your out-of-office message? Keep the message simple, brief, and clear. Here are a few examples (feel free to use them):

Example #1 (Straightforward)

"Thanks for your email. I'm away from my office and have limited access to my phone and computer. I'll reply when I return on September 17th."

Example #2 (An Alternate Contact)

"Thank you for your email. I'll be out of the office from July 7th through July 9th. I'll reply when I'm back at my desk. If this is an emergency, please email Sharon at sharon@acmecorp.com."

Example #3 (Fun And Honest)

"Thanks for your email. I'm currently on a digital detox to reclaim my life from my phone, email, and various gadgets. I'll get back to you on August 24th."

Example #4 (A Bit Of Snark)

"Thanks for your email. I'm currently at my favorite bar seeking enlightenment at the bottom of a Johnnie Walker bottle. I'll reply when I get back to my desk. Or maybe I won't. Deleting emails is easier when I'm drunk."

YOUR OUT-OF-OFFICE MESSAGE, whether it's straightforward, fun, or snarky, will reset others' expectations of when they'll hear back from you. That protects your reputation during your detox. They won't think you're being non-responsive out of laziness or spite.

PREP STEP 4: CREATE A LIST OF ACTIVITIES YOU'LL DO DURING YOUR DIGITAL DETOX

~

You've heard the saying "idle hands are the devil's workshop." That's certainly the case when it comes to your digital detox. You need to focus your attention on something other than your devices. Otherwise, the temptation to retrieve them may prove irresistible.

Plan activities to do while you're unplugged. For example, maybe you've wanted to improve your cooking skills. Now's the time to do it. Perhaps you've been aiming to plant a garden. That's a great way to spend your free time during your detox.

A lot of people use the time to reconnect with friends and family members. They plan lunch dates and enjoy the face-to-face interaction.

Whether you choose to improve your cooking, plant a

garden, or get together with loved ones, the important thing is that you have something to do.

Don't wait until you're without your phone and internet access to brainstorm activities. Do it beforehand. Write down a list of 10 projects that interest you. You don't have to do all of them. You just need to have the list available so you can pick and choose activities according to your mood.

You can take this idea one step further by scheduling activities during your detox. For example, make a lunch date with a friend and put it on your calendar. Call a sibling and pick a time to meet for breakfast. Set aside an hour to play cook, garden, play guitar, draw, or go hiking. Put it on your calendar and treat it like an appointment.

By filling your time with engaging activities, you'll be less inclined to surrender to the temptation to retrieve your gadgets.

PREP STEP 5: TELL YOUR FRIENDS AND FAMILY ABOUT YOUR DIGITAL DETOX

～

There are two reasons to tell the people in your life that you're planning to do a digital detox. First, it resets their expectations, a dynamic we discussed earlier. Second, it creates accountability.

People form expectations based on your actions. If you do something over and over, they'll expect you to do the same thing in the future.

For example, consider how quickly you respond to text messages. If you normally respond within seconds, the people you normally communicate with via text will always expect a speedy response from you. When you fail to text back in what they consider to be a "normal" time frame, they'll think something is wrong.

By telling your friends and family that you plan to spend time away from your devices, you force them to reset their expectations. We talked about this idea briefly in *Prep Step 3: Create An Out-Of-Office Email Message*. It has the same application here.

Telling your friends and family also makes you accountable. Behavioral psychologists have long known that telling others about our plans makes us more likely to follow through on them. We do so to avoid the shame we expect to endure from our friends and family members if we fail.

For example, recall the last time you declared your intention to break a bad habit (e.g. smoking, bingeing on junk food, etc.). Perhaps you told your spouse. Or if you were feeling adventurous, you posted it on Facebook.

Do you remember how the simple act of telling others motivated you to accomplish your goal? You didn't want to have to admit failure. And so you pushed yourself to do whatever you claimed you were going to do.

Leverage that same behavioral psychology to ensure you stick to your digital detox. Tell your friends and family about your plan. Post it on Facebook. Take advantage of the social pressure to stay unplugged. You'll find doing so makes it easier to resist the cravings for your favorite devices.

PREP STEP 6: MAKE A LIST OF POTENTIAL CHALLENGES YOU'LL FACE DURING YOUR DIGITAL DETOX

~

We discussed some of the obstacles you'll face in the chapter *What To Expect During And After Your Digital Detox*. It's important to write them down and keep the list in front of you. That way, you'll anticipate them and be able to address them in a productive manner when they surface.

For example, if you're addicted to your phone, you'll probably experience an effect known as phantom phone syndrome during your detox. That's where you think your phone is ringing or vibrating when, in reality, it isn't. If you anticipate experiencing this type of false alarm, you can take steps to manage - or at least ignore - it. Otherwise, it'll be a constant frustration.

By keeping a list of potential challenges in front of you,

you'll avoid being blindsided by them. Instead, you'll be aware of them before they occur. That gives you an opportunity to come up with compensatory strategies to deal with them.

PREP STEP 7: MAKE SURE YOU HAVE ACCESS TO IMPORTANT RESOURCES

~

We rely on the internet so much that we tend to take it for granted. We forget how prominent a role it plays in our daily experience. It's only when we're deprived of access - for example, when the power goes out in our homes - that we note its value.

This dependency can sabotage your digital detox. If you're like me, you've become so accustomed to having information at your fingertips that you rarely, if ever, write things down. After all, what would be the point? You can retrieve the information on your phone or laptop.

But during your detox, you won't have access to your devices. If you don't write down the info you'll need while you're unplugged, you'll be forced to make do without it.

Or you'll feel compelled to grab your phone, effectively breaking your detox.

Neither outcome is acceptable. Importantly, both are avoidable.

First, brainstorm the types of information you'll need during your detox. For example, are you planning to cook a meal according to a specific recipe? Do you intend to travel to a destination for which you'll need driving directions?

Once you've identified the information you'll need, look for it online. Print it out or write it down. That way, you'll have it at your fingertips and won't need your phone or laptop to look it up later.

Also, think about tasks you'll need to do before your detox begins in order to enjoy certain activities *during* your detox. For example, do you hope to visit a popular restaurant? If so, find the venue's phone number and make a reservation in advance. Do you intend to read a specific book on your Kindle while you're away from your other gadgets? If so, buy the book ahead of time so it will be waiting for you on your Kindle.

If you possess the resources you need, you'll be less inclined to succumb to cravings for your gadgets while you're on your digital retreat.

PREP STEP 8: COMMIT TO STICKING TO YOUR DIGITAL DETOX

～

Breaking an addiction requires commitment. Making a commitment prioritizes your efforts. It signals that you're willing to sacrifice something to effect a desired outcome.

In the case of a digital detox, making a commitment entails pledging to sacrifice your phone and other gadgets in order to sever your dependency on them.

Does committing to a project truly influence our behaviors and attitudes toward it? Research indicates that it does. In 2013, the journal Environment & Behavior published findings suggesting that making a commitment leads to long-term behavior change in most people. Other studies show that committing not only influences our behaviors and attitudes, but prompts us to act with greater consistency.

Before you start your digital detox, truly commit to it.

Review the reasons you're doing it (i.e. to break your addiction and dependency on your devices). Consider the challenges and obstacles you'll face (e.g. cravings, moodiness, inability to concentrate, etc.). Think about the ways in which your life will improve after you successfully complete your detox. You'll enjoy stronger relationships, more free time, and increased productivity.

Thinking about these aspects will help you to weigh the benefits you'll enjoy as a result of your detox versus the inconvenience of doing it. You need to have that balanced perspective in order to make a commitment to yourself.

Some people take the idea of committing to a digital detox one step further. They create a contract with a friend. The contract spells out the parameters of the detox and the consequences to be imposed if a relapse occurs. The penalty for giving in to cravings might take the form of a cash bond. For example, the addict agrees to pay his friend $50 or promises to make a donation to a cause or charity he detests.

The only limitation is the creativity of the involved parties.

You may be tempted to skip this step because you question its value. Don't underestimate the power of committing to your digital detox. It can mean the difference between successfully completing it and reaching for your phone and other devices when the first cravings surface.

PREP STEP 9: GET MOTIVATED AND INSPIRED!

∽

I n *Prep Step #8: Commit To Sticking To Your Digital Detox*, I recommended that you think about your reasons for doing a detox. That alone should give you the motivation and inspiration you need to complete it.

But there are other ways to get excited that will help you to maintain your enthusiasm throughout the ordeal. Here are a few ideas…

First, read about others' experiences doing a digital detox. FastCompany published one such account here (http://www.fastcompany.com/3012521/un-plug/baratunde-thurston-leaves-the-internet). The author, Baratunde Thurston, stepped away from the internet for 25 days and details the experience in his article. Business-Insider published another account here (http://www.busi-nessinsider.com/7-day-digital-detox-2016-3). The author,

Jeremy Berke, a 23-year old self-proclaimed phone addict, gave up his devices for a week at the behest of his boss. His insights about the time he spent unplugged are interesting (to say the least). Thurston's and Berke's narratives will inspire you regarding doing your own detox.

Second, deal with negative thoughts the moment they surface. Whenever we try to overcome an addiction or break a bad habit, a little voice in our heads makes defeatist comments. Sometimes the voice is subtle. It gently suggests we'd enjoy ourselves more by feeding the habit we're trying to break.

Other times, it's aggressive. It tries to convince us that we'll never be able to successfully break our addiction. Ergo, trying to do so via unplugging from our devices is a waste of time and effort.

Such negative thoughts will drain your energy. That, in turn, will erode your motivation to complete your digital detox. It's important to deal with such thoughts directly to prevent them from gaining a foothold in your mind.

When the voice in your head makes cynical comments, banish them immediately. Don't consider them. Don't debate them. Don't negotiate. Simply drive them away.

A third way to get inspired about your detox is to meditate on it. This suggestion might sound strange. If so, you may have a false impression of what meditation is.

You don't need to sit on the floor, cross your legs, and chant with incense burning in the background. You don't need complete silence. You don't need to close your eyes and concentrate on an empowering image.

At its core, meditation is nothing more than thinking about something with focus. It's contemplation. If you've ever taken a leisurely walk and spent the time mulling something over, you've already done it.

Do it for your digital detox. Imagine how you'll feel during it. Think about the challenges you'll face and visualize overcoming them. You'll be surprised by how just a few minutes of quiet contemplation can inspire you to succeed!

This is the last of the nine steps in preparation for your digital detox. Everything we've covered thus far in this action guide has been in anticipation of the next section. You're now ready to set aside your gadgets and reap the benefits of a life untethered to technology.

PART II

10 STEPS TO DOING A COMPLETE DIGITAL DETOX

~

Giving up your phone, tablet, video game console, and internet access is going to be a shock to your system. Your brain has become accustomed to the dopamine rush it experiences whenever you feed your addiction. It won't want to forfeit that sensation. Expect your brain to revolt, triggering intense cravings for the things you've temporarily given up.

The key to a successful digital detox is to create an environment that minimizes the chances of a relapse. You need to insulate yourself so you can withstand the temptations to give in.

That's what the following 10 steps accomplish. Each

one is tasked with severing your connection to your devices and the internet.

Let's start with your phone…

STEP 1: BURY YOUR PHONE

∼

Turning off your phone notifications isn't enough. Nor is turning off your phone altogether. You need to put it somewhere out of sight, and hopefully out of mind.

If you keep your phone near you during your digital detox, you'll be tempted to use it. It's human nature. If you're addicted to something, your brain will do everything it can to compel you to act on your compulsions.

It starts with rationalization (*"How much will it really hurt to look at your phone just this one time?"*). It progresses to making defeatist thoughts (*"You're going to fail, so you may as well look at your phone."*). Finally, it manages to wear down the addict and convinces him to feed his habit.

Don't underestimate your brain's ability to persuade you to do what it desires. It's crafty that way. It knows you're obsessed with your phone and will use every tactic at its disposal to convince you to use it.

The solution is to bury your phone (not literally). Throw it into a dresser drawer. Place it at the back of your bedroom closet. Give it to your spouse. The most important thing is that prevented from seeing it.

Out of sight, out of mind.

What if you use your phone as an alarm clock in the morning? Should you make an exception and leave it on your nightstand for that purpose?

Absolutely not. It will be too great a temptation.

Buy a cheap alarm clock to use during your detox. A small travel model shouldn't cost more than $10. It will allow you to put your phone out of sight, helping you to resist the inevitable cravings you'll experience.

STEP 2: HIDE YOUR TABLET

∿

I f you own an iPad, Samsung Galaxy Tab, Microsoft Surface Pro, or any other tablet, you own a device designed for consuming digital media.

Sure, it's possible to use it productively. You can send emails on it. You can create spreadsheets. You can even write a book (though you'll want to use an external keyboard for that task). But the majority of your time will be spent watching videos, playing games, visiting Facebook, and reading articles, blogs, and other written content.

In other words, you'll use it to consume various types of media. That's what tablets are made for.

And that's the reason you need to hide your tablet along with your phone.

It's easier said than done. There's a good chance you use your iPad or Surface Pro throughout the day. You might always have it at your side, whether you're working in your office, lounging on your couch, or (gasp) visiting the bathroom.

That's why it needs to be out of sight during your detox. Your constant use has probably contributed to your technology addiction. Like any drug, you should have zero access to it while you're unplugged.

STEP 3: STORE YOUR LAPTOP

∼

I live on my Macbook Air. I use it for everything, from checking email and writing blog posts to working on client projects, planning out goals, and creating action guides like the one you're currently reading.

Although I have a smartphone, it's not my lifeline. That role is filled by my laptop.

I mention this because if I don't store my laptop away when I'm unplugged, I'm going to open it. It's a foregone conclusion. So it's not enough for me to tell myself to resist the temptation to use it. If it's near me, I'm going to fire it up.

That's the reason I strongly encourage you to store your laptop out of sight when you're doing your personal detox. Put it with your phone and tablet in a closet. Or give it to a loved one and make her promise to not surrender it no matter how much you beg or threaten her.

If this is your first detox, this step is crucial. Why? Because your first experience will set your expectations for every subsequent detox you perform. Get the first one right, and you'll be more inclined to stick to the ones you do in the future. Fail your first time, and your brain will learn that failure is an option - and an appealing one to boot!

STEP 4: UNPLUG YOUR COMPUTER

∾

I can't walk past a computer without wanting to jump on the keyboard. It's a weakness. As I mentioned in *Step #3: Store Your Laptop*, I live my life on the computer. If my house were burning down, the first things I'd grab are my Macbook Air and external hard drive (after making sure my wife was safe, of course). Nothing else comes close in terms of priority.

In the old days, I used to work on a desktop computer. Even back then, I was obsessed. I'd find myself gravitating toward it regardless of what I was doing in that particular moment.

There was always a purpose. For example, I'd look something up online, update a spreadsheet, or write part of an article or book. The point is, I was addicted. There were times during which I was cooking and I'd run to my home office to hop on the computer for the 60 seconds I had at my disposal.

I had zero impulse control.

So here's my advice: if you use a desktop computer, unplug it before starting your digital detox. Shut it down. Completely. Don't just put it in sleep mode. Turn it off. Otherwise, it will pose too great a temptation when you're trying to stay unplugged.

I speak from experience.

STEP 5: DISCONNECT YOUR OFFICE COMPUTER FROM THE INTERNET

~

The ideal time to do a digital detox is on the weekend or while you're taking a vacation. You'll be able to unplug from your phone and the internet without concern about tasks related to your job. You'll be able to relax and fill your time with offline activities.

That said, you may have a job that requires your attention seven days a week. It might entail spending a part of your weekends at the office. The idea of taking a vacation may be laughable given your responsibilities.

If that's the case, you might be forced to do your detox in the office. It's not an ideal setting, but manageable with the right approach. To make it work, you'll need to commit to avoiding the internet.

Therein lies a problem. As long as your office computer

remains connected to the internet, you'll face the temptation to feed your technology addiction. You'll be tempted
to check your email; you'll long to check out Facebook,
Twitter, and Pinterest; you'll want to visit CNN, YouTube,
and your favorite blogs.

Remember, you're fighting an addiction. Your brain
will be looking for ways to compel you to relapse.
Willpower alone won't be enough to help you resist the
temptations and remain true to your detox.

Think of it this way: you wouldn't expect a cocaine
addict to resist the temptation to use while carrying a vial
of coke in his pocket, right? By carrying the vial, he's
setting himself up for failure.

So it is with working on a computer connected to the
internet while on your digital detox.

With that in mind, sever the connection. Turn off your
Wi-Fi. Get as much done as possible without internet
access. When you need to check your email, reconnect and
do so quickly. Then, sever the connection again.

If you must work in the office, you're already at a
disadvantage. Your access to the internet will challenge
your ability to successfully complete your detox. Each time
your brain tells you to visit Facebook, catch up on the latest
news headlines, or watch cat videos on YouTube, you'll
know you're just a click away. The temptation to falter will
be nearly impossible to resist.

The best way to minimize the problem is to turn off
your connection to the internet.

STEP 6: WEAR A WATCH

~

Twenty years ago, if you wanted to know the time, you'd look at your wristwatch. Or you'd ask someone nearby who was wearing one.

Those days are largely gone. More people than ever use their phones to check the time than look at their watches. In fact, few of us even wear watches anymore. When we do, they're more of a fashion accessory than anything else.

If you're addicted to your phone, this poses a problem. You'll need a way to find out the time during your digital detox, but can't risk carrying your phone with you. The temptation to use it would be too great.

This article in Wired Magazine (http://www.wired.-com/2015/08/bought-dumb-watch-rescue-phone) describes the effect. The author writes that he mostly uses his phone to check the time. But once he does so, he's tempted to use it for other time-wasting activities, such as checking out the latest posts on Facebook.

And down the rabbit hole he goes.

This impulse is overpowering for the tech addict. For that reason, I strongly recommend that you wear a watch during your detox. That way, you can find out the time without exposing yourself to the siren call of your phone.

If you don't own a watch, buy one. It needn't be expensive. You can pick up a cheap Casio digital watch at Amazon for less than $12. It won't look pretty, but it'll do the job.

If you're concerned the watch will clash with your outfit - after all, ditching technology doesn't mean ditching your fashion sense - remove the wristband and stick the part with the digital display into your pocket or purse.

The important point is that you have a way to tell the time that doesn't require your phone or other devices. If you live life with one eye on the clock, this will be a major step in successfully completing your digital detox.

As I mentioned earlier, it's also a good idea to buy an alarm clock. If you're like most people, you rely on your phone to rouse you from sleep in the morning. Since you won't have access to your phone - remember, you've hidden it from sight or given it to a friend for safekeeping during your detox - you'll need another option. A reliable travel alarm clock will cost less than $15 at Amazon.

STEP 7: SCHEDULE "OFFLINE" ACTIVITIES WITH FRIENDS AND FAMILY

~

I n *Prep Step #4: Create A List Of Activities You'll Do During Your Digital Detox*, we talked about the importance of having things to do. Things to fill your time. If you become bored, you'll be tempted to grab your phone, pull out your video gaming console, or hop onto the internet.

Remember, "idle hands are the devil's workshop."

This step is so important that it's worth revisiting in the context of doing your detox. It's one thing to come up with a list of activities you'll focus on while you're unplugged. It's another thing entirely to put those activities on your calendar so you're left with minimal downtime.

First, call your friends and family members. Plan coffee dates, lunch dates, dinner dates. Put the dates and times on your calendar. Research shows we're more likely to follow through on something if it's scheduled. We naturally assign the activity a high priority.

Second, schedule blocks of time during which you can pursue your interests. For example, suppose you enjoy playing tennis. It's not enough to tell yourself that you're going to do more of it during your digital detox. Call a friend and ask him or her to join you at a specific time. Put it on your schedule. You'll not only reduce your idle time, but you'll have something concrete to look forward to.

Third, if you've been wanting to visit a particular locale, whether it's the beach, a new museum, or a swanky jazz club, now's the time to do it. Again, put it on your calendar. Remember, that which gets scheduled gets done.

Being unplugged is difficult when you're battling a phone or internet addiction. You need a way to keep your mind occupied. Enjoying "offline" activities with your loved ones will give you an opportunity to have fun in the absence of your gadgets.

STEP 8: MAKE CONNECTIONS WITH STRANGERS

~

Most of us shy away from meeting strangers. Whether we're introverts or simply too busy or not in the mood for unfamiliar company, we tend to stick to ourselves when we're alone.

Our phones and other devices are enablers in this respect. They give us something to focus our attention on in situations when we're in public, but trying to avoid making eye contact with others. Ever watched people waiting for drinks at Starbucks? Most of them will stand near the bar and fool around with their phones rather than talking to the person standing next to them.

When you're on a digital detox, you won't have your phone to fall back on in social situations. That's a positive thing. It will spur you to come out of your shell and connect with people face to face. Take advantage of it!

When you're unplugged, you'll find that it's easier to talk to people. You may even look forward to doing so despite having never taken the opportunity in the past. Real connections, even if created through a few moments of small talk, are deeply gratifying to us.

Try this on your next digital detox: strike up a conversation with a stranger. If you're at a coffee shop, lean over to the person sitting next to you and comment on the shop's food. If you're standing in line at the grocery store, turn to the individual standing behind you and give them a sincere compliment. If you're taking a walk and cross paths with someone unfamiliar to you, greet them and comment about the weather. If he or she is walking a dog, make a flattering remark about it.

Most strangers you meet will share some of your thoughts on a myriad of subjects. It's just a matter of tapping into them. It's a great way to take your mind off your phone, your computer, and other gadgets, and enjoy life unplugged.

STEP 9: READ A BOOK

∾

Our phones and other gadgets have diverted time and attention away from one of most rewarding activities we can experience: reading a good book.

Many people love to read, but assume they lack the time. More accurately, they spend their available time consuming other types of media. Little is left over to spend reading a rousing romance novel, a gripping fantasy tale, or an absorbing scientific fiction yarn.

When you do your digital detox, you'll have plenty of time to read a book (or two). Take the opportunity to do so. Immerse yourself in a story that takes your mind off your gadgets. Lose yourself in a narrative that captivates you and draws you to its end.

Reading books while unplugged not only keeps your

mind occupied. It also improves your imagination, refines your ability to focus, and lowers your stress. Scientists have also found that habitual reading leads to a better memory, more empathy, and even improved sleep.

So there's lots to gain by picking up a book during your detox.

Now, a few tips…

First, don't use your phone or tablet to read books. You'll be tempted to log onto Facebook, watch YouTube videos, and play games. Instead, use your Kindle - not a Kindle Fire, though, since they're designed to encourage you to consume digital media - or buy a hard copy.

Second, lean toward full novels rather than short stories. Remember, you want an immersive reading experience. You want to get to know the characters to the point that you care about them. To that end, you want a story that takes those characters through a satisfying arc during which they experience failure, growth, and change. A short story rarely delivers on those points as successfully as a good novel.

Third, read fiction. You may be drawn to the latest release on investing strategies, exercise and dieting, or how the brain works. But such books won't do during your digital detox. You need a book that expands your imagination and boosts your creativity. You need one that pulls you in and occupies your mind with a powerful story. A nonfiction book may be interesting, informative, and insightful, but it won't tap into the creative region of your brain.

Pick a few novels to read during your detox. If you're

unsure which ones to buy, I recommend looking through Amazon's lists of bestsellers (http://www.amazon.com/best-sellers-books-Amazon/zgbs/books). Drill down to a section that interests you - for example, *Mystery, Thriller & Suspense* - and pick a few that catch your eye.

STEP 10: LEARN A NEW SKILL

~

Preparing a list of activities to do while you're on your digital retreat is important. As we discussed earlier, the activities will keep your mind occupied, making you less susceptible to temptations to use your gadgets. And trust me, the temptations *will* come.

I also recommend learning a new skill. It will allow you to focus your attention on the pursuit of something that will help you to grow as a person.

Sure, going on lunch dates, visiting the beach, going bowling, tending to your garden, and enjoying a spa day, are great ideas for things to do during a digital detox. But these activities, while fun, relaxing, and mood-enhancing, are primarily for personal gratification.

When you learn a new skill, you're devoting yourself to something that poses lasting value. It broadens your repertoire of abilities. You'll learn how to do something that was previously outside your skill set.

Pick something you've been interested in pursuing, but have put on the back burner due to lack of time and motivation.

Need a few ideas to get the creative juices flowing? Learn how to:

- Play the guitar
- Speak Italian
- Change the oil in your car
- Repair a faucet in your home
- Defend yourself against an assailant
- Cook Japanese cuisine
- Code a website
- Take engaging photographs
- Create spreadsheet graphs and charts
- Speedread
- Analyze stocks
- Practice yoga
- Give a public speech
- Apply first aid
- Perform karate
- Swing dance
- Create a flower arrangement

The above is just the tip of the iceberg, of course. The goal is to find a new skill that interests you and learn to do it competently during your digital retreat.

Immerse yourself in the activity with the intention of mastering it.

You can also pick a topic in which to become more knowledgeable. For example, suppose you've always wanted to learn about astronomy. Why not read a book about it while you're unplugged? Or let's say you're interested in knowing more about the history of Rome. Grab a book about it and keep it near you during your detox.

Here are a few other subjects that will serve as a springboard for your own ideas and interests:

- Psychology
- Tort law
- Thermodynamics
- Controlled burns
- Game theory
- Anthropology
- Macroeconomics
- Endocrinology
- Car transmissions
- Primate behavior
- Surgical procedures
- Philosophy
- Biographies of famous people
- Socratic method
- Theology
- Autism
- Nutrition

Again, the point is to pick a subject that interests you. Then, devote some of your "unplugged" time to learning more about it. Schedule the sessions on your calendar.

You'll find that learning a new skill or becoming more knowledgeable about an unfamiliar subject will keep boredom at bay during your detox. As a bonus, it will also improve your ability to concentrate.

PART III

THE EFFECTS OF A DIGITAL DETOX ON YOUR BRAIN

∾

You're doing this for a reason. Why else would you set aside the gadgets you love? There must be a benefit, something that will make your life more rewarding, waiting for you at the end of your digital detox.

Here's the good news: the time you spend on your technology sabbatical will introduce *several* benefits, each of which can have a profound effect on your life.

We'll cover nine of these benefits in the remaining pages of this action guide. What follows isn't meant to be an exhaustive list of the advantages you'll reap from setting aside your devices. Instead, I'll highlight the ones that are likely to have the greatest impact on your daily life.

Let's start by taking a look at the effect a digital detox will have on your productivity.

EFFECT #1: IMPROVED PRODUCTIVITY

~

Your phone, tablet, video game console, and the internet, are distractions. Whether you're at home or working in the office, they make it more difficult for you to get things done.

Consider a typical workday. How often have you tried to complete a project or task, but been distracted by text messages, emails, Facebook updates, and YouTube videos? That's not even counting the impromptu interruptions from your family (if you're at home) and coworkers (if you're in the office).

Scientists claim the brain requires 20 minutes to regain its momentum following each interruption. That's the reason it often seems like time is slipping through your fingers!

A digital detox removes the tech-related distractions

that are hampering your productivity. You're able to focus on whatever project or task is in front of you. This allows you to achieve a flow state, where your productivity can double, and even triple.

After you've completed your detox, you'll find that your productivity will remain higher than it was prior to setting aside your gadgets. That's because your detox will have gone a long way toward severing your dependency on technology. It's an important step toward breaking your addiction and controlling your impulses. You'll be more inclined to use your phone and other devices to help you get things done rather than letting them usurp control of your life.

If you haven't gone on a technology sabbatical, your devices probably seem indispensable. For example, texting allows you to stay in contact with your friends, which helps you to maintain relationships that are important to you. Your Xbox or PS4 help you to relax, which reduces stress. And how else will you stay in touch with your internet friends unless you check in to Facebook each day?

But all of that is an illusion. There are healthier ways to accomplish these things without feeding your tech addiction and allowing your productivity to plummet.

I'm certain you'll agree after you've experienced the peacefulness that results from unplugging from your devices and the internet.

EFFECT #2: STRONGER RELATIONSHIPS

∾

I deally, technology would improve the relationships we share with the people who are important to us. Software should make it easier to plan get-togethers. Coordinating busy schedules should be a snap.

In reality, our phones and constant access to the internet tend to do the opposite. We neglect our family members and friends, even when they're right in front of us. Instead, we compulsively check our phones, looking for texts and emails that validate us and indicate others are thinking about us.

The ramifications are sobering.

In 2012, the Journal of Social and Personal Relationships published a study examining the impact of mobile devices on the quality of face-to-face conversation. The authors found that the mere *presence* of a phone interferes with the intimacy and trust shared between two people.

With that in mind, one of the most rewarding benefits of doing a digital detox is that you'll enjoy stronger relationships. You'll be more present with friends and loved ones, giving them the attention they deserve. In return, you'll earn their trust and empathy, which will reinforce the bonds you share with them.

Prior to your detox, you may have compulsively checked your phone for new messages or social media updates, even when having lunch with someone. After taking an extended technology break, you'll find it easier to resist the compulsion.

It's a good tradeoff. Instead of being glued to your phone while a friend tries to have a meaningful conversation with you, you'll be ready to truly connect with him or her. You'll signal your interest in what he or she is saying through eye contact and active listening. That's the path to stronger, more gratifying relationships.

EFFECT #3: REDUCED STRESS

∼

S tress is ever-present. There's no way to avoid it. Few people go through an entire day without feeling its effects, even if they fail to realize it.

Short bursts of stress can be useful. They help us to respond to conflict and external pressures (the "fight or flight" response). The problem is, many people experience *persistent* stress. They develop a chronic form of anxiety that prevents them from relaxing.

The consequences are so severe that an entire industry has sprung up in recent years to help people manage their long-term stress levels. Unfortunately, our addiction to technology is rarely seen as a contributing factor.

We talked about stress in the chapter *Top 12 Side Effects Of Technology Addiction*. Compulsive behavior spurred by the addiction puts undue stress on both the mind and body.

Over an extended period, that can lead to worrisome health impacts.

A digital detox brings instant relief. You'll spend time away from your gadgets, giving your brain a much-needed rest. Rather than trying to keep up with text messages, emails, and social media updates, along with your daily work-related responsibilities, you can relax. You can think clearly and be mindful of your surroundings. You can regain a healthy perspective concerning the things in your life that are important to you.

Lowering your stress levels also poses health benefits that extend beyond your current state. Scientists claim it will reduce your susceptibility to heart trouble, protect you from migraines, aid digestion, and bolster your immune system. Doctors claim it can even improve your sex drive.

Some psychologists opine that stress is at the root of our happiness, or lack thereof. They suggest that true contentment can only occur if we find a way to manage our stress levels in a positive way.

Again, stress is a fact of life. It will always be so. But that doesn't mean you should tolerate or endure persistently high stress levels.

The good news is that doing a digital detox can cause much of the stress you're feeling to evaporate.

EFFECT #4: BETTER SLEEP

∾

If you're like most phone addicts, you compulsively check for new texts, emails, and Facebook updates throughout the day. And chances are, you even do it in bed right before you go to sleep.

This obsession hampers your sleep in at least two ways.

First, you end up going to sleep later because you're busy checking for new content. Assuming you have to get up in the morning at a specific time, this limits the amount of rest you receive each night.

Second, staring at your phone reduces the volume of melatonin in your body. (I mentioned this effect, caused by the blue light emitted by your phone's screen, in the chapter *Top 12 Side Effects Of Technology Addiction*.) Melatonin influences your sleep and wake cycles. You need a proper amount to enjoy a night of restful slumber.

As you go through a digital detox, you'll notice that the quality of your sleep improves. This is because you'll have resolved the two problems outlined above. You're unplugged from your phone, and therefore won't be tempted to stay up late checking for new messages. As a result, you'll get more sleep.

Additionally, being unplugged prevents you from spending hours staring at your phone's blue-light screen. The level of melatonin in your body will no longer be suppressed, which means the hours you spend asleep will be more restful.

This isn't just about sleeping better at night and thereby feeling better in the morning. Improved sleep delivers a variety of benefits, many of which influence your long-term health. You'll enjoy a better memory, more creativity, sharper focus, and less stress. You'll also experience better weight control, less susceptibility to heart problems, and a stronger immunity system.

To recap, being addicted to your phone, tablet, or the internet is likely having a negative impact on the amount and quality of your sleep. That's the case for a large percentage of technology addicts. The good news is that a digital detox - taking a much-needed sabbatical from technology - can remove the obstacles preventing you from enjoying a full night's rest.

EFFECT #5: LONGER ATTENTION SPAN

∾

I mentioned earlier, back in the chapter *Top 12 Side Effects Of Technology Addiction*, that today's average attention span is less than that of a goldfish. Our phones, video games, social media sites, and YouTube, are largely responsible.

We're bombarded with a continuous stream of notifications. And each time one arrives, we're compelled to check its source. Did a friend just post an update on Facebook? Did I just receive a text message from my significant other? Did someone just email me about something I'm interested in?

Each notification distracts us. It interrupts what we're doing and breaks our concentration. It's no wonder goldfish are beating us in terms of attention span!

One of the first things you'll notice after completing your digital detox is that you're able to focus. You'll realize that few notifications pose emergencies that need your immediate attention. As such, you'll be less inclined to jump at your phone each time it buzzes or chirps.

The joy of being unplugged is that you sever the around-the-clock connectedness that feeds your technology addiction. You cut off the source of your habit.

Imagine the effect that will have on your attention span and ability to concentrate. You'll no longer be distracted by a never-ending string of notifications that pull you away from the task at hand. You won't feel compelled to reach for your phone every few minutes "just in case" you missed a text message, email, or Facebook update.

If you're in the office, you'll get more work done. You'll find it easier to enter a flow state, where you're not only able to focus, but you're also excited and energized by whatever you're working on.

If you're relaxing with your family, you'll be more present with them. Your spouse and kids will notice that you're paying more attention to what they have to say. You'll become a more active listener rather than being constantly distracted by your phone.

It's easy to downplay the advantages of having a longer attention span. After all, we live in a society that places a high value on the ability to multitask while being inundated with interruptions. But productivity experts have long known that it's through focused attention that we're able to gain true control over our lives.

That control gives us clarity of purpose. It helps us to make good decisions. It allows us to resolve challenges in a way that accommodates our values and goals.

And that's the path to living a rewarding life.

EFFECT #6: GREATER SELF-DISCIPLINE

∽

At the core of every addiction, whether it's to technology, cocaine, or sex, is an inability to resist the compulsion to act. Unfortunately, each time we act to satisfy the compulsion, we reinforce the addiction. We covered this concept earlier, but it's important enough to repeat here.

Self-discipline alone isn't enough to beat an addiction. The brain's reward system becomes "rewired" as it develops a dependency on the object of obsession. Once that happens, no amount of willpower in the world will keep the compulsion to act at bay.

But once an addiction has been beaten, after one's dependency has been shattered, self-discipline becomes paramount during recovery. It plays a critical role in the addict's ability to set limits on what she allows into her life.

These limits don't prevent her from enjoying a rewarding lifestyle. They do the opposite. They prevent her from falling again into the snare of addiction. They prevent a relapse.

After you complete your digital detox, you'll experience a greater level of self-discipline. You'll know instinctively that reaching for your phone or checking Facebook whenever the mood strikes is inconsistent with your goals. You'll be aware that the compulsion to do so conflicts with your values, and over the long run will hamper your life rather than make it better.

Most importantly, you'll have the courage to resist the impulse to feed your habit. Your detox will have shown you that it's possible to withstand the siren call of technology.

Will it be difficult? Yes. Especially in the beginning, and particularly after your first detox.

But each time you resist the urge to reach for your phone, it becomes easier to do so. Each small victory over your addiction, no matter how seemingly inconsequential, will give you more confidence that you can control your impulses.

After your second and third digital detoxes - it's a good idea to plan a few tech sabbaticals each year - you'll find it easier still.

That's evidence of greater self-discipline. It's like a muscle. The more you exert it, the stronger it grows.

EFFECT #7: IMPROVED CREATIVITY

~

C reativity is the ability to come up with original ideas. It's about having a fertile imagination.

Experts claim technology hampers our creativity. They note that while our computers, phones, and other gadgets help us to get more done (a dubious claim in and of itself), they're eroding our ability to innovate. Our reliance on technology is slowly diminishing our originality.

This effect stems, in large part, from our exposure to tech-related distractions. Text messages, emails, social media updates, video games, YouTube videos, and the ever-changing headlines that reflect today's minute-by-minute news cycle demand our attention. In doing so, they disrupt our workflow and destroy our concentration.

It's difficult to be creative when you're unable to focus.

It's hard to come up with inventive ideas when you're bombarded with digital media every minute of your day.

When you do a digital detox, you free yourself from that situation. Unplugged from your gadgets and disconnected from the internet, you're free to think deeply about ideas and concepts. You're able to ponder them in a way that's impossible when you're constantly distracted.

You're also free to be bored. Occasional boredom is good. It spurs creative thought. A study published in the Journal of Experimental Social Psychology in 2014 showed that bored participants performed better than engaged participants on a variety of creativity tests.

A fertile imagination introduces benefits that extend beyond the ability to come up with new ideas. You'll be a better problem solver. You'll be more self-aware. And according to researchers, you'll experience less stress. Improved creativity will also lead to more meaningful relationships, increased spontaneity, and greater self-confidence in your personal values and vision.

Following your first digital detox, don't be surprised if you experience a level of creative thought that is unfamiliar to you. And further, don't be surprised if your newfound creativity leads to a greater sense of personal fulfillment. It's a natural effect of quarantining yourself from the onslaught of digital media.

EFFECT #8: BETTER INTERPERSONAL COMMUNICATION

～

I n the chapter *15 Ways Your Life Will Improve After A Digital Detox*, I noted that your social skills will improve once you unplug from your gadgets. This effect is largely because you won't be distracted by them. You won't be inclined to grab your phone in the middle of a conversation just because you've received a new text message. You won't feel the compulsion to immediately go onto the internet to validate a small, but unimportant, piece of information in the middle of a friendly chat.

Without such distractions, you'll be better able to focus on the person you're speaking to. You'll be more tuned in to their nonverbal cues. You'll become a better listener, and as a result a better communicator.

Most of us crave personal interaction with others. We enjoy talking to people, whether they're friends or strangers. To that end, good interpersonal skills accord numerous benefits, all of which make life more rewarding.

First, you're able to express your thoughts with greater clarity. Your intuition about the other person's reception of your message sharpens, giving you an opportunity to tailor your message to his or her interests.

Second, you enjoy stronger relationships, both at home and at the office. Listening to people builds trust and empathy, the two key building blocks of any relationship.

Third, you become a more effective leader. Being a strong, clear communicator inspires others. Your message has greater conviction, which increases your audience's confidence in you.

Fourth, good interpersonal skills make you a better team player. You're able to work more effectively in groups, earning other members' trust, confidence, and loyalty.

Fifth, you become a better negotiator. You'll be more adept at moving discussions forward with the goal of creating deals that allow all parties to benefit.

Technology addiction impairs our ability to communicate and connect with others. Our phones and other devices are obstacles to enjoying meaningful conversations and building strong relationships.

Once you complete your first digital detox, you'll notice a marked improvement in your ability to relate to others. That change will lead to the many benefits that come with developing solid interpersonal skills.

EFFECT #9: STRONGER MEMORY

~

A growing body of research shows that our compulsion to check our phones, surf the internet, and log onto social media, is affecting our recall. According to scientists, these activities impair our short-term - or working - memory capacity.

In the chapter *How Technology Addiction Negatively Impacts Your Life*, I mentioned a 2011 study that appeared in the journal PLoS One. It described the findings of researchers in China studying the effects of the internet on the brain. The researchers found that significant exposure caused "structural changes" in the gray matter. They noted these changes could disrupt the brain's memory functions.

Studies also show we're less inclined to commit things to memory when we believe we can retrieve them online. In fact, we habitually record details solely to avoid the chore of remembering them. This practice is what drives the success of online tools like Evernote, Google Keep, and Microsoft OneNote.

An article in Wired Magazine in 2010 (http://www.wired.com/2010/05/ff_nicholas_carr/) had the following to say:

"Dozens of studies by psychologists, neurobiologists, and educators point to the same conclusion: When we go online, we enter an environment that promotes cursory reading, hurried and distracted thinking, and superficial learning. Even as the Internet grants us easy access to vast amounts of information, it is turning us into shallower thinkers, literally changing the structure of our brain."

Researchers also note that short-term memory is delicate. It's fragile. Any distraction can wipe it clean. That's one of the reasons technology addiction, with its nonstop distractions, is hampering our ability to remember things.

Spending time away from your phone and the internet removes the digital distractions. It allows your brain to retain information in its short-term memory, and pass important details to its long-term memory.

An improved memory introduces a variety of benefits that make life more rewarding. You'll find it easier to learn new skills, grasp new concepts, and resolve personal challenges. You'll also enjoy better recall of events and details, making you a better conversationalist. Ultimately, a strong memory makes you a well-rounded person.

PART IV

AFTER YOUR DIGITAL DETOX

AFTER YOUR DIGITAL DETOX: 10 THINGS TO DO TO AVOID A RELAPSE

∿

Before we discuss what you should do after completing your first digital detox, allow me to congratulate you! Going without your phone, laptop, and other devices - even if only for 24 hours - isn't easy. Living unplugged can make it seem as you're missing out on a vital part of your daily experience. You'll feel incomplete without access to the internet. You may feel naked without the heft of your phone in your pocket.

So consider your detox to be a triumph. It's a major victory in finally beating your addiction to technology. It should be celebrated!

Having said that, the work doesn't end here. Controlling your compulsions to feed your habit is an ongoing process. It's similar to the alcoholic who must remain ever vigilant lest he relapse and succumb to the siren call of the bottle. You must remain alert and watchful lest you once again fall under technology's spell.

Below, I'll highlight the 10 things you should do after completing your first digital detox. Many of them require ongoing maintenance and diligence. The upside is that they'll enrich your life through improved productivity, stronger interpersonal relationships, and the abundance of other life benefits we covered in previous chapters.

#1 - Maintain Phone-Free Conversations

No one *needs* to look at their phone while having a conversation. When people do it, they do so out of compulsion. They lack impulse control.

One of the best things you can do to prevent your phone from overtaking your life is to put it away during conversations. Don't put it on the table near you. Turn it off and store it out of sight.

Focus on the person you're with. Give him or her your full attention. I guarantee you'll enjoy a more meaningful conversation and experience a stronger connection with that individual.

As a bonus, you'll become better able to control the urge to check your phone for new messages.

#2 - Clean Up Your Browser Bookmarks

Do you know how many bookmarks are stored in your favorite browser? If you're like me, you've stockpiled thousands over the years and use less than 1% of them.

The majority are a distraction. They pull your attention away from whatever deserves your focus. You might even feel overwhelmed by them to the point that you're unable to act.

That kills your productivity when you're trying to research things on the internet.

The solution is to purge most of your bookmarks. Review each one and try to remember the last time you used it. (I use the Chrome browser. The easiest way to review your bookmarks in Chrome is to open the Bookmark Manager.) If you haven't used a bookmark in the last 12 months, there's little chance you'll ever use it. So get rid of it.

I go through this exercise once a year, getting rid of entire folders at one time. I recommend that you do the same.

#3 - Declutter Your Phone

Your phone is the ultimate distraction. With it, you can reach out to any contact, read any web page, and play any game (depending on platform - iPhone or Android) whenever the mood strikes. You can watch YouTube videos, read news headlines, and check into Facebook, Twitter, and Pinterest, on a whim.

It's no wonder so many people have difficulty getting things done!

Here's my advice: declutter your phone. First, get rid of the apps you no longer use. Then, get rid of those you know to be a complete waste of time. Next, put your photos and videos in the cloud and delete them from your phone. Lastly, purge files you no longer need. That includes old podcasts and music you haven't listened to in months.

Think like a minimalist. Remove the clutter to weaken your phone's hold on you.

#4 - Limit The Number Of Open Browser Tabs To Seven

Tell me if this sounds familiar… you're reading an article online and come across a link to another article that looks interesting. You open it in a new tab so it will be there waiting for you when you finish reading the first one.

At that moment, you have a random thought about a recipe you'd like to try for dinner that evening. You open a new browser tab to search for it on Google.

With your focus broken, you think to yourself, "I should check my email." You open another tab to do so.

After checking your email, you return to the first article, which you haven't finished reading because you're distracted. While reading it, you stumble upon an unfamiliar concept. You decide to research it immediately lest you forget to do so later. You open Google in a new tab, and within minutes have opened five more tabs to investigate the concept.

Before long, you have dozens of tabs open. And you know you'll never get to them all.

Here's my advice: commit to having no more than seven tabs open at one time. Seven is manageable. You'll be able to see at a glance what each tab contains as opposed to only seeing favicons (the tiny images displayed in the tabs) due to a lack of screen space. You'll also avoid the distraction and feeling of overwhelm that accompany having tons of tabs open.

My record is 42 tabs. I'm not bragging. Having that many tabs open destroyed my productivity. Learn from my mistake.

#5 - Protect Your Email Inbox

From the moment you finish your digital detox, you should consider your email inbox to be sacred ground. Few things are more distracting than an inbox littered with dozens of unread messages. Your goal should be to keep the number of such emails to a minimum.

A lot of people advocate a practice known as "Inbox Zero." This is an email management strategy that aims to eliminate every message from your inbox.

Personally, I think it's unnecessary. I'd even argue it's a waste of time since the advantages of having an empty inbox are outweighed by the time and effort required to keep it so.

Instead, I recommend doing the following:

- Unsubscribe from newsletters you haven't read for the past three months.
- Unsubscribe from "shopping deal" emails that haven't spurred you to buy anything in the last month.
- Ask family and friends to refrain from sending you unnecessary messages (e.g. jokes, links to YouTube videos, etc.)
- Archive or delete old messages. If you choose to archive them, create folders and labels to streamline the process of retrieving them in the future.

These four steps will keep your email inbox clutter to a minimum. Your inbox doesn't need to be empty. It just needs to be manageable so you can avoid wasting time on unnecessary (or old) messages.

#6 - Renounce "Drive-By" Google Searches

Remember the days when, while having a conversation with a friend, you'd disagree about an unimportant tidbit of information? Back then, you might have argued in good fun. But ultimately, if you were unable to convince your friend that you were correct, you'd agree to disagree.

It was the only way to resolve the matter and remain on good terms.

Today, such disagreements can be resolved on the spot. If there's ever a question about something, you can launch an impromptu fact-finding mission by accessing Google on your phone. The answers are literally at your fingertips.

The problem is, every time you reach for your phone to perform a "drive-by" Google search, you reinforce the habit of doing so. The practice, repeated over and over, slowly destroys your impulse control.

Try this: the next time you're wondering about a piece of information, resist the urge to look it up on your phone. Instead, realize that, barring an emergency or work-related issue, you probably don't need to know the answer.

In fact, *not* knowing the answer poses its own benefits. It may prompt you to ponder the subject more deeply, asking questions to gain greater insight. That process, in turn, will improve your critical thinking skills.

#7 - Reduce Your Digital Media Consumption

Think about the different types of media you're bombarded with every day. Blogs, articles, YouTube videos, Facebook updates, Twitter tweets, product reviews, photo galleries, animated gifs, webinars, and the list goes on. You can literally spend every waking moment consuming this media.

And that's not even including watching your favorite

television shows on Netflix!

After you complete your first digital detox, review the types of media you consume on a regular basis. Write down each type along with the trigger that drives you to it.

For example, do you watch YouTube videos whenever you're bored? Do you check in to Facebook as a way to procrastinate? Do you binge-watch Netflix shows whenever you're stressed?

Next, identify activities that can replace your consumption of digital media as a response to the triggers. For example, if you're feeling stressed, go for a walk instead of turning to Netflix. If you're bored, play with your dog instead of watching YouTube videos (your dog will love the attention).

The goal is to limit the amount of media you consume each day. By doing so, you'll weaken the hold your technology addiction has over you, and improve your impulse control in the process.

#8 - Purge Unneeded Facebook "Friends"

Do you really need 1,000 Facebook friends? Are they truly friends you trust with the most intimate and embarrassing details of your life? Or can you live without them?

If you're like most people, you rarely connect with the majority of the folks you're friended on Facebook. They show up in your newsfeed, but you're rarely compelled to reach out to them. You might "like" one of their photos or leave a comment on something they've posted on their page. But at the end of the day, you have little in common with them.

They're not friends as much as acquaintances. And they're barely that.

The trouble with having a ton of Facebook "friends" is that they're a distraction that reinforces your tech addiction. You end up watching your newsfeed for new posts and comments, spurred by the fear of missing out.

I recommend you cut through your friends list with a digital scythe. Keep only those whom you consider to be true friends. Purge the rest.

Facebook makes this easy. This page (https://www.facebook.com/friends/organize) will list the "friends" with whom you haven't interacted in awhile. Go through the list and unfriend them.

You can always like their pages and receive updates that way.

#9 - Turn Off Phone Notifications

Phone notifications are a terrible feature for two reasons. First, they're a continuous distraction. At work, it's impossible to maintain any modicum of flow and momentum when your phone is constantly chirping.

It's no better at home or when you're out with friends. Have you ever tried to have a conversation with someone when his or her phone is chiming, beeping, and lighting up? It's an object lesson in patience!

The second reason phone notifications are a terrible

feature is because they wear down your impulse control. Each time your phone chirps, you instinctively reach for it to learn the reason. Did someone send you a text message or email? Did a friend just update her Facebook page? Did a new version of an app download to your phone?

When you respond to every notification, you train yourself to do so down the road. The act becomes a habit. It feeds your compulsion and reinforces your addiction. In the end, you become like Pavlov's dog, salivating at the sound of a bell.

Turn off your phone notifications. You don't need them. In the event of an emergency, you'll likely receive a phone call if you don't respond immediately to a text or email.

#10 - Establish Personal Rules

The surest way to maintain your technology "sobriety" after your digital detox is to set ground rules concerning how you'll use your devices and the internet going forward. Establish guidelines that dictate when you'll use them, for how long, and in what situations. Following are several examples with respect to non-work related use:

Your phone - use it in the morning before work, during your lunch break, and from 6:00 p.m. to 8:00 p.m. Never use it outside of those times. Never do so in the company of others. Limit your use to 15-minute sessions.

The internet - use it from 6:00 p.m. to 7:00 p.m. If you

need to look up something urgent, do so on your lunch break. Don't surf aimlessly; always have a purpose.

YouTube - watch videos from 6:00 p.m. to 7:00 p.m. Don't watch out of boredom and without purpose. Search for specific videos you want to watch. For example, search for "Steve Jobs Commencement Speech" or "how to braise short ribs."

Social media - check in no more than once a day. Do it on your lunch break. Limit your session to 15 minutes.

Video games - play games from 9:00 p.m. to 9:30 p.m. Limit yourself to three times a week.

These are merely examples. The guidelines you set for yourself should reflect the areas you struggle with the most. For instance, I don't play video games, so I don't need to set a personal guideline for it. My weakness is the internet. I have to be proactive in limiting my use.

The Road Forward

Again, congratulations for completing your first digital detox! You've taken a crucial step toward curbing your addiction to technology. It's a step most tech addicts will never take. The upside is that you'll enjoy a host of life-enhancing benefits that will remain out of their reach.

But once is not enough.

Plan to do a digital detox two or three times each year. Why? Because addictions never truly disappear. They're suppressed and kept under control, but they're always there, waiting for an opportunity to regain a foothold in your life. Doing a periodic detox will keep the impulses at bay.

You don't need to abandon your phone and the internet for weeks on end. A 24-hour detox every four to six months should suffice. Or do it more often if you feel your self-control is slipping.

You're at the helm. You're the one calling the shots. You determine how often to unplug. The most important thing to recognize is that taking periodic breaks from your gadgets will improve your life in myriad ways.

It will ground you and help you to develop impulse control. It will allow you to reconnect with others, strengthening the relationships that are important to you. It will introduce numerous cognitive and physiological health benefits. And it will help you to suppress your technology addiction.

Ultimately, doing periodic digital detoxes will set the stage for enjoying a more rewarding, engaging lifestyle.

MAY I ASK YOU A SMALL FAVOR?

∾

I'd like to take a moment to thank you for reading this action guide. It means the world to me. I'm an independent author, so one of my biggest priorities is to spread the word about my books.

If you enjoyed reading *Digital Detox: The Ultimate Guide To Beating Technology Addiction, Cultivating Mindfulness, and Enjoying More Creativity, Inspiration, And Balance In Your Life!*, would you consider letting others know? There are lots of ways to do that. You could:

1. Leave a review on Amazon (click here to do so)
2. Leave a review at Goodreads.com
3. Tell folks about it on your blog
4. Share it on Facebook or Twitter

Positive reviews encourage others to give independent authors like me a chance. They help loads. And believe me, I can use all the help I get.

Thanks again for taking the time to read *Digital Detox*. I sincerely appreciate it.

If you'd like to be notified when I release new action guides (typically at a discount), please sign up for my mailing list at http://artofproductivity.com/free-gift/. You'll receive a free PDF copy of my 40-page guide *Catapult Your Productivity: The Top 10 Habits You Must Develop To Get More Things Done*. You'll also receive periodic tips, tricks, and hacks for managing your time and designing an enriching, fully-satisfying lifestyle.

ALL THE BEST,

Damon Zahariades
http://artofproductivity.com

OTHER BOOKS BY DAMON ZAHARIADES

The Mental Toughness Handbook

The definitive, step-by-step guide to developing mental toughness! Exercises included!

To-Do List Formula

Finally! Discover how to create to-do lists that work!

The Art Of Saying NO

Are you fed up with people taking you for granted? Learn how to set boundaries, stand your ground, and inspire others' respect in the process!

The Procrastination Cure

Discover how to take quick action, make fast decisions, and finally overcome your inner procrastinator!

Fast Focus

Here's a proven system that'll help you to ignore distractions, develop laser-sharp focus, and skyrocket your productivity!

The 30-Day Productivity Plan

Need a daily action plan to boost your productivity? This 30-

day guide is the solution to your time management woes!

The 30-Day Productivity Plan - VOLUME II

30 MORE bad habits that are sabotaging your time management - and how to overcome them one day at a time!

The Time Chunking Method

It's one of the most popular time management strategies used today. Triple your productivity with this easy 10-step system.

80/20 Your Life!

Achieve more, create more, and enjoy more success. How to get more done with less effort and change your life in the process!

Small Habits Revolution

Change your habits to transform your life. Use this simple, effective strategy for adopting any new habit you desire!

Morning Makeover

Imagine waking up excited, energized, and full of self-confidence. Here's how to create morning routines that lead to explosive success!

The Joy Of Imperfection

Finally beat perfectionism, silence your inner critic, and overcome your fear of failure!

The P.R.I.M.E.R. Goal Setting Method

An elegant 6-step system for achieving extraordinary results in every area of your life!

For a complete list, please visit

http://artofproductivity.com/my-books/

ABOUT THE AUTHOR

~

Damon Zahariades is a corporate refugee who endured years of unnecessary meetings, drive-by chats with coworkers, and a distraction-laden work environment before striking out on his own. Today, in addition to being the author of a growing catalog of time management and productivity books, he's the showrunner for the productivity blog ArtofProductivity.com.

In his spare time, he shows off his copywriting chops by powering the content marketing campaigns used by today's growing businesses to attract customers.

Damon lives in Southern California with his beautiful, supportive wife and their frisky dog. He's currently staring down the barrel of his 50th birthday.

www.artofproductivity.com